T0205919

The Rise of the Intelligent Health System

"I recommend that all members of the health community read this book to obtain a real snapshot of how the Intelligent Health System is being transformed via new technologies."

–Chris Landon *MD FAAP, FCCP, FRSM, Clinical Associate Professor USC Keck School of Medicine, Technology Development Center Laboratory and Studio*

The "Intelligent Health Pavilion" as demonstrated at the annual HIMSS Conference by the Intelligent Health Association is the impetus for this book.

This book documents the remarkable journey of "Intelligent Health System" and the adoption of Innovative technologies. Many showcased in real time on the trade show floor and now in this book: "The Rise of the Intelligent Health System". In each chapter of this book, authors are expressing the immense potential of merging cutting-edge technology with the complex realm of patient care and safety. The informative chapters in this book delve deep into the unfolding story of how hospitals have evolved into interactive, intelligent environments, driven by real-time data and powered by artificial intelligence.

In what seems like the blink of an eye, technology has completely transformed the way we live, work, and interact with the world around us. From smartphones to self-driving cars, ChatGPT, wireless technologies, wearables, and many other innovations are reshaping our society, pushing the boundaries of what was once considered impossible. However, nowhere is the impact of technology more profound than in the field of healthcare.

Intelligent Health Series

Leveraging Technology as a Response to the COVID Pandemic
Paul H. Frisch, PhD and Harry P. Pappas

Innovating Healthcare with Wearable Patient Monitoring
Mike Davis, Michael Kirwan, Walt Maclay and Harry P. Pappas
2023

The Rise of the Intelligent Health System
Paul H. Frisch & Harry P. Pappas
2024

The Rise of the Intelligent Health System

Paul H. Frisch, PhD
and Harry P. Pappas

First published 2024
by Routledge
605 Third Avenue, New York, NY 10158

and by Routledge
4 Park Square, Milton Park, Abingdon, Oxon, OX14 4RN

Routledge is an imprint of the Taylor & Francis Group, an informa business

ISBN: 978-0-367-76935-2 (hbk)
ISBN: 978-0-367-76934-5 (pbk)
ISBN: 978-1-032-69031-5 (ebk)

DOI: 10.4324/9781032690315

Typeset in Garamond
by MPS Limited, Dehradun

Dedicated to my wife, Susan, and our children and their spouses, Paul & Katie, Lauren & Tom, and Daniel & Gabby and our grandchildren, Reagan, Mia, Austin, Zoey, Landon, and Collins, who provided me the inspiration to contribute to the work highlighted in this book. Looking toward the future, my wish is that the continued advancement of healthcare technologies will have a significant positive impact on the health, wellness, and longevity of my family and all families across the world.

–Paul H. Frisch, PhD

Dedicated to my wife, Linda, and to my children: Mark Pappas and Maria Pappas Sparling; my daughter in law: Giorgia Mesiti Pappas; my son-in law Steve Sparling; along with my grandchildren, Penelope and Giada Pappas. I also would like to thank and dedicate this book to the hundreds of thousands of dedicated healthcare workers around the world who treated and saved the lives of thousands of Coronavirus patients during this horrible pandemic of 2020–2022.

Not only did these heroic people spend countless hours helping to treat and save lives but they also put their own lives at risk and that of their families to HELP OTHERS.

We would also be remiss if we did not thank the many scientists and researchers in the pharmaceutical industry and the academic community for their unstinting efforts and dedication to conducting the exhausting research needed to discover a vaccine that would vanquish this deadly virus. Without these dedicated healthcare professionals, the Pandemic could have been worse.

Thank you for your Herculean efforts to make the world a bit safer for all.

–Harry P. Pappas

Founder and CEO, The Intelligent Health Association

Contents

Foreword ... ix

About the Intelligent Health Series ... x

About the Authors ... xi

About the Contributors ... xiv

1 The Rise of the Intelligent Health Consumer ... 1

TOM LAWRY

2 Fundamentals of the Intelligent Health System ... 12

PAUL H. FRISCH

3 Visibility In Healthcare with IoHT ... 38

FAWZI BEHMANN

4 Medical Device Security Program for a Healthcare Delivery Organization ... 52

ALI YOUSSEF

5 Wearable Devices and Remote Patient Monitoring ... 73

WALT MACLAY

6 Explosion of Robotics in Healthcare ... 87

ROGER SMITH

7 RFID in Healthcare ... 112

SHAAN REVURU

8 Developing an Institutional RFID–RTLS Strategy and Management Plan 143
PAUL H. FRISCH

9 The Role of Real-Time Location Systems in Ambulatory Care 162
JOANNA WYGANOWSKI AND MARY JAGIM

10 Informatics and Analytics 171
JAMES BEINLICH

11 Intelligent Healthcare Use of Germicidal Ultraviolet "C" (UVC) Light 187
ARTHUR KREITENBERG

12 Optimizing Infection Control and Hand Hygiene 202
MATUS KNOBLICH

13 Hospital at Home: Transformation of an Old Model with Digital Technology 213
ALISA L. NIKSCH

14 The Digital OR 230
GREG CARESSI AND BEJOY DANIEL

15 Techquity 245
CHRIS LANDON

Appendix I: Glossary of Terms 250

Index 264

Foreword

In what seems like the blink of an eye, technology has completely transformed the way we live, work, and interact with the world around us. From smartphones to self-driving cars, innovations have reshaped our society, pushing the boundaries of what was once considered impossible. However, nowhere is the impact of technology more profound than in the field of healthcare.

The rise of the intelligent hospital system represents a remarkable journey, one showcased in this book, with each expert expressing the immense potential of merging cutting-edge technology with the complex realm of patient care. This book delves deep into the unfolding story of how hospitals have evolved into interactive, intelligent environments, driven by data and powered by artificial intelligence.

Through the pages that follow, readers will be taken on a captivating exploration of the myriad ways in which technology is revolutionizing healthcare. From electronic medical records that streamline patient information to robotic assistants that support medical procedures, the authors expertly guide us through the exciting landscape of the intelligent hospital system.

As the reader will see through these expert's eyes, knowing and understanding the benefits of this ongoing transformation are essential.

Chris Landon MD FAAP, FCCP, FRSM
Clinical Associate Professor USC Keck School of Medicine
Technology Development Center Laboratory and Studio

About the Intelligent Health Book Series

The Intelligent Health Association (IHA) is pleased to be publishing a series of educational books under the series name of the: "Intelligent Health Series" in cooperation with our publisher, Taylor and Francis, a div. of Informa. This will be one of many books in the series.

The IHA advisory board and the IHA Educational Committee have identified experts in many areas of health and wellness and have formed several committees, comprised of healthcare professionals and technology thought leaders for the purpose of publishing a set of technology-centric books for the health and wellness community.

This book will be targeted to clinicians and health and wellness professionals who are focused on exploiting diverse, new evolving technologies to enhance patient care and optimize the workflows and processes of patient, triage, diagnosis, treatment, and care.

We organized a team of healthcare professionals and experts to both author and edit chapters in this book. The publication of this book and others will be available online through Routledge Taylor and Francis Group Content from the books will also be repurposed into audiobooks, podcasts, webinars, and conferences in cooperation with its editors and authors.

These and other books in the Intelligent Health Series will be marketed at future conferences promoted by IHA and their partners, through Wikipedia, LinkedIn, and various industry and training events worldwide where IHA participates.

Let's educate the global health and wellness community together.

Thank you

Harry P. Pappas, Founder and CEO
Dr. Paul Frisch, Pres. IHA
Intelligent Health Association

About the Authors

Paul H. Frisch PhD, FHIMSS
An Attending, Department of Medical Physics
Director of Biomedical Engineering
Memorial Sloan-Kettering Cancer Center
New York, NY

Paul Frisch is currently an Attending in the Department of Medical Physics and the Director of Biomedical Engineering at Memorial Sloan-Kettering Cancer Center. He currently serves on technical advisory boards of IEEE Counsel on RFID, ECRI Institute, and serves as the Chief Technical Officer for the RFID in Healthcare Consortium and President of the Intelligent Health Association.

Paul's responsibility focuses on the investigation of new and evolving technologies and their potential integration into clinical applications and operations to enhance patient outcomes, care, and safety. This includes management of Technology and Medical equipment to ensure device integrity, patient safety, and regulatory compliance. Current areas of investigation focus on clinical 3D imaging and printing, robotics, medical device development, and cybersecurity.

Previous experiences include research in electromagnetic field-induced gene expression focused on targeted gene therapy, human–robotic application in pharmaceutical development, and biodynamic response resulting from transitory acceleration, such as crash impact and aircraft ejection.

Paul Frisch has a Doctoral degree in Biomedical Engineering from the State University of New York at Binghamton (2008) and Master's and Bachelor's degrees in Electrical Engineering from the State University of New York at Stony Brook (1975, 1976).

Harry P. Pappas

Harry Pappas is a successful, High Tech., Serial entrepreneur with a strong focus on the health technology sector. He is a strong believer in applying technology to transform the health and wellness community in today's "Continuum of care," from the hospital, to the primary caregiver, and to the patient's Smart Home. Pappas firmly believes that the world of digital health is being driven by the adoption of technology and therefore the need for Quality, ON-GOING education.

Harry is a global, thought leader and has been a tech geek since the age of 12. He is a speaker at many Health and Wellness conferences and trade shows around the world.

He and his team are the producers of the award winning "Intelligent Health Pavilion™" a technology-centric, DIGITAL HOSPITAL, that you may have visited at many trade shows around the world, including at HIMSS over the last TEN years.

Pappas is the Founder and CEO of the Intelligent Health Association, a global, Social Purpose entity, educational, technology-centric organization dedicated to helping educate members of the Healthcare community on the adoption of new technologies. These technologies can improve patient care, patient outcomes, and patient safety, while driving down the cost of healthcare.

Pappas is the creator of the (i-HOME™) a health and wellness SMART HOME concept that demonstrates, "In Context" a plethora of health and wellness technologies placed in a Digital Smart Home setting for Remote Patient Monitoring (RPM) and Patient Management. Harry was developing the concept of the "Smart Home" utilizing Steve Jobs', original Apple "NEWTON" PDA device many, many years ago.

Harry is an internationally recognized thought leader with auto-ID, BLE, NFC, RFID, RTLS, Sensors, Voice, Robotics, Wearables, AI, P 5 G, Biomarkers, AI, and Wireless technologies. He has been presenting educational programs around the world since 2001.

Harry is also an Industry Ambassador in the Open Voice Network (OVON), www.openvoicenetwork.org, an affiliated organization of the Linux Foundation. Pappas chairs the Health, Wellness, and Life Sciences Community of the OVON.

He is clearly an "out of the box thinker" and a long-term strategic player in the world of health technologies for the Digital hospital and for today's Smart Home.

Harry is unique in that he has "hands-on" experience with a wide variety of technology and software development projects as they relate to the health and wellness industries.

Harry's Goal: To help educate the healthcare community on an going bases, so that it may adopt new technologies, software, APPS, Voice, 5 G, Blockchain, and AI that can have a dramatic impact on the delivery of improved health.

Mantra: "Help Others", Do "SOCIAL GOOD"

About the Contributors

Fawzi Behmann, DL, MBA, M. Comp. Sc., Author
President, TelNet Management Consulting Inc.
IEEE ComSoc North America Board Director and BoG member
Co-founder of "IoT in Healthcare Consortium TM" of the Intelligent Health Association (IHA)

Fawzi is currently the founder and president of TelNet Management Consulting Inc. offering international professional services in the areas of technology trends; positioning and building smart networking ecosystem solutions in key markets. Prior, Fawzi held various executive and leadership positions with Tier 1 companies in the areas of communications and networking in Canada and the USA. He championed the development of Telecom Network Management systems and led efforts in rolling out product releases for network edge and core and marketing SoC product lines and roadmap.

Fawzi has been a keynote speaker and distinguished lecturer at several domestic and international conferences and events. He is the Co-founder of "IoT in Healthcare Consortium TM" of the Intelligent Health Association (IHA). He has several publications and co-authored a book on "Collaborative Internet of Things for Future Smart Connected Life and Business" published by Wiley.

Fawzi volunteered at IEEE and he is currently ComSoc North America Regional Director and BoG member. Fawzi championed the delivery of several conferences and summits and has been the general chair for IEEE WCNC 2022 held in Austin (April 10–13). He is a voting member of the IEEE Conference Committee and Co-Chair of IEEE SA Transdisciplinary Framework working groups and a contributor to IEEE Future Direction Initiatives. He chairs multiple joint chapters on Communications, signal processing,

Computer, EMBS, and Consumer technology. Prior, Fawzi was the IEEE Region 5 conference committee chair and Central Texas section chair.

Fawzi received multiple awards including IEEE Communications Society NAB Exceptional Service Award (2020 and 2021); IEEE Leadership Section chair for Central Texas Section (2017–2018); IEEE USA professional leadership award for 2017; ComSoc Chapter Achievement Award (2021); IEEE ComSoc Chapter of the Year and Chapter Achievement Awards (2020, 2017, 2015); MGA Regional outstanding Member Award (2013, 2014 and 2015); and Freescale CEO Diamond Chip Award (2008).

Fawzi has an MBA, a Masters in Computer Science, and a Bachelor in Science with honors in Mathematics with distinction.

James Beinlich

James Beinlich's experience includes large, complex health systems in the areas of data and analytics, project and program management, strategic planning, process redesign, IoT, and operations improvement. Jim has consulted for large private health systems and academic medical centers as well as the US Department of Defense and the National Institutes of Health. He has held numerous senior leadership positions in large healthcare systems and was most recently the Chief Data Information Officer at a large academic health system. Jim currently has adjunct faculty appointments at Temple University's College of Public Health and Widener University's Graduate School of Business.

Greg Caressi

Senior Vice President, Global Client Leader
Healthcare & Life Sciences
Frost & Sullivan

Greg Caressi is a business leader and healthcare industry expert with more than 25 years of experience in new technology assessment, business case development, competitive analysis, and geographic expansion. Mr. Caressi is responsible for Frost & Sullivan's Healthcare and Life Sciences practice, with a focus ranging from digital health to medtech to next-generation life science solutions.

Mr. Caressi engages with industry leaders across the globe regarding their technology needs and gaps: healthcare providers, payors, life science research and commercial organizations, medical technology vendors, and individual consumers.

Mr. Caressi is a frequent conference speaker and consultant to leading companies in the digital health sector. He led the HIMSS Life Sciences IT Committee as Chair for two years, creating a set of published Good Informatics Practices for pharma/biotech companies.

Prior to Frost & Sullivan, Greg worked in economic development and export promotion with the Yunnan Finance & Trade Institute (Kunming, China), Mabei Taili Group Co. (Jiangsu, China), and Northern California District Export Council (San Francisco, CA).

Mr. Caressi earned an MBA with emphasis in Pacific Basin Studies from Dominican University, a Bachelor of Arts in Economics from Miami University, and a Bachelor of Science in Education from Miami University.

Dr. Bejoy Daniel
Senior Industry Analyst
Healthcare & Life Sciences
Frost & Sullivan

Dr. Bejoy Daniel has over 23 years of experience in the healthcare industry, comprising work across market research, strategy consulting, business analysis with global healthcare and medical device companies, R&D in healthcare IT and medical devices, and implementation of primary healthcare models in rural and semi-urban settings.

As a senior member of Frost & Sullivan's Healthcare & Life Sciences team, Dr. Daniel produces customized analysis and identifies future growth opportunities in the healthcare industry, with a focus on:

- Digital Models for Future Hospital Operating Rooms
- Emerging Business Models in Drug Delivery Devices
- Changing Paradigms in Orthopedic and Spinal Surgery Devices
- Global Robot-assisted Surgical Devices
- Clinical Decision Support Solutions for the Intensive Care Unit
- Global Surgical Navigation Systems

Dr. Daniel earned a MBL with specialization in M&A, Investment and Institutional Finance from the National University of Juridical Sciences and NUJS, and a Bachelor of Dental Surgery (B.D.S.) from Dr. M.G.R. Medical

University. Dr. Daniel also completed the Dasra Social Impact Leadership Program with Harvard Business Publishing

Mary Jagim, MSN, RN, CEN, FAEN | Principal Consultant at CenTrak

Mary Jagim is an experienced leader in healthcare consulting with an expertise in real-time location systems, emergency nursing, healthcare operations, and public policy. In 2001, she served as the national president of the Emergency Nurses Association (ENA) and led the development of ENA's Key Concepts in ED Management Course and Guidelines in Emergency Department Staffing Tool. She was also a member of the Institute of Medicine (IOM) Study on the Future of Emergency Care and the National Quality Forum ED Consensus Standards Committee. As the Principal Consultant for CenTrak, Mary works with healthcare organizations to leverage real-time technologies coupled with process enhancements to improve the patient experience, patient and staff safety, and workflow efficiency. Having implemented hundreds of real-time location system projects in the last 15 years, Mary is one of the most experienced clinical leaders and implementers in Healthcare RTLS in the world and developed the "Jagim Lean RTLS Model for Healthcare." She also currently serves as a member of the IoT Community, Healthcare Advisory Board.

Matus Knoblich

Matus Knoblich was born in former Czechoslovakia, a political refugee who escaped communism with his family in 1987. Growing up in New York and Florida, he attended the University of Virginia where he attained a Bachelor of Science degree in Chemistry with a special focus in Biochemistry. Matus has focused on global business development, sales, and marketing since 2005, with an international assignment in Geneva, Switzerland for 6 years, conducting business in over 100 countries. During this time, Matus focused on company restructuring and building. In 2016, he assumed management of Med-Stat Consulting Services Inc., a medical device service company with a specific focus on hospital bed repair. At this time, he also founded Glo-Med Networks Inc., with a focus on sales and distribution of new and novel medical devices and consumables. In 2019, Glo-Med began global distribution of the Orbel personal hand sanitizer, with many new and exciting products about to enter distribution as development is completed in the coming months and years. Matus opened a European office in Switzerland in 2020 to support Europe, Middle East, and Africa via founding of Glo-Med Networks AG. Matus has continued his work in the sector, opening support businesses such as medical logistics facilities in the greater New York area and western Florida, along with a transport company, counting some of the largest and most respected healthcare facilities in the USA as his clients. With a devotion to family and staff, interests in snowboarding and boating, Matus looks to the future of healthcare and how he can have a direct impact for positive change.

Arthur Kreitenberg, MD, FACS

Dr Arthur Kreitenberg is a Board Certified Orthopedic Surgeon and award-winning Clinical Professor at the University of California, Irvine School of Medicine. He has published numerous peer-reviewed studies and has served as a reviewer for peer-reviewed journals. His interest in UVC disinfection technology dates back to 2009 H1N1 outbreak when he developed a device to disinfect volleyballs and basketballs that was used by Team USA in the 2010 London games. He co-founded Dimer UV and serves as the Chief Innovation and Technical Officer and has developed UVC products for aviation, spaceflight, farming, healthcare, education, athletics, and building environments. The GermFalcon for aviation was named the best travel innovation in 2018 by the BBC and by the Global Business Travel Association. This UVC aircraft disinfecting robot, together with the GermRover, a robotic zerogravity drone for disinfecting spacecraft, received top honors at a NASA iTech competition. In 2020, Dimer entered into a strategic partnership with Honeywell to commercialize the GermFalcon. The UVHammer for healthcare has won industry awards for its novel design that overcomes inherent challenges of UV exposure including distance, shadowing, and angles of incidence. Dr. Kreitenberg and Dr. Martinello of Yale University co-authored a paper recommending tough standards for UVC devices to protect patients. He is a member of the Association of Professionals in Infection Control and the American Institute of Aeronautics and Astronautics, an Associate Fellow of the Aerospace Medical Association, and a Fellow of the American List of Contributors ▪ xxiii Academy of Orthopedic Surgeons. He was a two-time NASA Astronaut Selection finalist.

Chris Landon, M.D.

Dr. Landon has directed large population-based investigations, patient recruitment, and physician education in COPD (CMS Innovation Grant), pediatric asthma (AAP), rheumatic heart disease, multiple sclerosis (Genentech), and cystic fibrosis. Current use of telemedicine includes opioid use disorder (USDA Distance Learning and Telemedicine), machine learning in COPD, and transition of adult cystic fibrosis patients from a Cerner to Epic EHR (CF Foundation).

Chris Landon, M.D., received his B.S. in psychobiology from the University of California, and his medical degree from the University of Southern California. He was a pediatric intern and resident at Stanford University Hospital and received additional training at Stanford's Children's Hospital. He is board certified in both pediatrics and pediatric pulmonology and is the director of pediatrics at Ventura County Medical Center. He is also a clinical assistant professor of family medicine at UCLA and of pediatrics at USC. He has contributed to articles in the *Journal of the American Medical Association*, the *Journal of the American Board of Family Practice*, and other professional publications. He is the director of the Pediatric Diagnostic Center in Ventura, California.

Tom Lawry

Tom Lawry is a leading AI transformation advisor to health and medical leaders around the world, a top keynote speaker, and the best-selling author of *Hacking Healthcare – How AI and the Intelligent Health Revolution Will Reboot an Ailing System.*

He's the Managing Director of Second Century Tech and a former Microsoft exec who served as National Director for AI for Health and Life Sciences, Director of Worldwide Health, and Director of Organizational Performance for the company's first health incubator. Prior to Microsoft, Tom was a Senior Director at GE Healthcare, the founder of two venture-backed healthcare software companies, and a health system executive.

Tom's work and views have been featured in Forbes, CEO Magazine, Harvard Business Review, CNET, Inside Precision Medicine, and numerous webcasts and podcasts.

In a Harris Poll of 2023 JP Morgan Healthcare Conference attendees, Tom was named one of the most recognized leaders driving change and engagement in healthcare today. He has also been named one of the Top 20 AI Voices to Watch.

Representative Keynote Presentations

"Intelligent Aging is Healthcare's Moonshot," Dublin Global Longevity Summit, Dublin, Ireland, August 2023

"Finding You AI Superpowers – Making the Intelligent Health Revolution Real," Vitalis 2023, Gothenburg Sweden, May 2023

"Step Aside Status Quo – The Intelligent Health Revolution is Here," American Telemedicine Association, San Antonio, March 2023

"The Future of Healthcare is Not What it Used to Be," Xccelerate 2023, Palo Alto, March 2023

"The Intelligent Health Revolution is Upon Us – Are You Ready?" HealthFront, NYC, April 2022

Representative Articles/Media Coverage

Why AI is Set to Become an Important Healthcare Tool, not a Threat, CEO Magazine, May, 2023.

How AI And Machine Learning Will Impact the Future of Healthcare, Forbes, September 2022

Hacking Healthcare—How AI and the Intelligence Revolution will Reboot an Ailing System, Inside Precision Medicine, October 2022

Walt Maclay

Mr. Walt Maclay, President and founder of Strawberry Tree inc, dba Voler Systems, one of the top electronic design firms in Silicon Valley, is committed to delivering quality electronic products on time and on budget. It provides design, development, risk assessment, and verification of new devices for medical, consumer, and industrial applications. Voler is particularly experienced in designing wearable and IoT devices, using its skill with sensors and wireless technology. The company has developed hundreds of products including medical devices, wearable devices, home health, products for the aging, and other medical, consumer, and industrial devices.

Mr. Maclay is recognized as a domain expert in Silicon Valley technical consulting associations. He has spoken at dozens of events on sensors, wearable devices, wireless communication, security for medical devices, and low-power design. In 2019, he wrote Highly Successful Engineering Design Projects, a book about project management available from Amazon. From 2008 to 2010, he was President of the Professional and Technical Consultants Association (PATCA). He is a senior life member of the Institute of Electrical and Electronic Engineers (IEEE) and a member of the Consultants Network of Silicon Valley. He has been an instructor at Foothill College in the Product Realization Certificate Program, teaching successful new product introduction skills. He has applied his outstanding leadership to many multidisciplinary teams that have delivered quality electronic devices. Mr. Maclay holds a BSEE degree in Electrical Engineering from Syracuse University.

Mr. Maclay is active in helping technology startup companies. He has participated in angel investor groups and has advised dozens of startup companies on technical and funding issues. He has mentored startups at Techstars and Cleantech Open. He is a reviewer for NSF SBIR grants. As the founder of his own company, he has dealt with issues of funding, product development, marketing, sales, and finance, giving him the experience to advise others. Voler Systems is a member of a technology consortium, the Product Realization Group, which provides all the services to introduce new technology products. Mr. Maclay started and led two forums for CEOs of small companies to discuss issues of importance to them.

Voler clients, from early-stage startups to large established corporations, are delighted with the quality and service they have received. Past clients include Alexza Pharmaceuticals, Applied Materials, BAE Systems, Boeing, Boston Scientific, Intel, JDS Uniphase, Life Technologies, Lockheed Martin, Maxon Lift, Merck, NASA Ames, Northrop Grumman, Orbital Sciences, Puget Sound Naval Shipyard, Radiant Medical, Rain Bird, Sandia Labs, Siemens, Spectra Physics, St. Jude Medical, Stanford University, Teikoku Pharma, Teradyne, Thoratec, the University of California at Davis, Zosano Pharma, and hundreds of others.

Alisa L. Niksch, M.D

Dr. Alisa Niksch is a pediatric cardiologist and electrophysiologist and currently serves as Senior Director of Medical Affairs at Owlet Baby Care, Inc. She has lent her experience to the digital health, medical device, and remote patient monitoring fields since starting her practice at Tufts Medical Center in 2010. Dr. Niksch was Chief Medical Officer of Genetesis, Inc., a company which created a novel cloud-connected and AI-powered cardiac diagnostics and imaging platform, where she led pivotal clinical trials and managed medical affairs initiatives. She has been an advisor and researcher for multiple healthcare companies like AliveCor, Cohere Health, Ometri, PraxSim VR, Medaica, Mindchild Medical, Zephyr Technologies, and Sproutling. She continues to be a startup mentor with programs at Northeastern University and MassChallenge HealthTech. She has authored a book chapter and articles in peer-reviewed journals on digital health and a wearable technologies and has spoken on the applications of AI in medicine and the role and design of wearable technologies in clinical practice. She is a graduate of The University of Virginia School of Medicine and completed her cardiology and electrophysiology fellowship training at Morgan Stanley Children's Hospital at Columbia University Medical Center and Stanford/UCSF Medical Centers, respectively.

Shaan Revuru

Shaan Revuru is a serial entrepreneur with a passion for creating value, generating growth, and delivering sustainable business and technology solutions that have a positive impact on people's lives around the world. Shaan is the founder of several startups including, FaceSpotR Corporation, CoolShopR Inc., OpenAppliance Corporation, ManageTheCrowd Inc., and Content OverDrive Inc. delivering innovative solutions. Shaan is currently the Chief Operating Officer at Pycube, a healthcare-focused technology company. Shaan has worked in leadership and management roles at Tetheron, Oracle, Ernst & Young, and Baan (now Infor).

Shaan has dual MBA degrees from Georgetown University – Robert E. McDonough School of Business & Edmund A. Walsh School of Foreign Service and ESADE Business School, an MS in Technology Management from George Washington University – School of Business and an undergraduate degree in Accounting and Economics from University of Delhi.

Dr Roger Smith

Dr. Roger Smith is an award-winning expert in robotic surgery training, education, and simulation. He is a Faculty Scholar at the University of Central Florida's College of Medicine and the Institute for Simulation and Training. He has led organizations in healthcare, defense, and education, serving as an executive Chief Technology Officer for AdventHealth Hospital Systems; US Army Simulation; and the Titan Division of L3Harris. He holds a PhD in Computer Science and a Doctorate in Management and has received service awards from AdventHealth, the US Army, Association for Computing Machinery, Society for Computer Simulation, and National Training and Simulation Association.

Joanna Wyganowski, MBA, PMP | Senior Director of Commercial Marketing at CenTrak

Joanna Wyganowski works with progressive healthcare organizations to leverage Real-Time Location Solutions (RTLS) to reduce costs and improve patient and staff safety and experience. She has been sharing RTLS best practices through the RTLS in Healthcare Community. Joanna is a certified Project Management Professional, Lean Master, and a UAV Pilot.

Ali Youssef

Ali Youssef is the Wireless Chair of the IHA and the director of medical device and IoT security at Henry Ford Health. He is an HIMSS Fellow with 22+ years of experience and holds several industry certifications including CPHIMS, PMP, CISSP, HCISPP, CISM, and CWNE. He has authored several articles on topics ranging from digital health to network architecture and medical device security. Ali authored a book entitled "Wi-Fi Enabled Healthcare" in 2014. He sits on the AAMI Healthcare Technology Leadership Committee and the AAMI Editorial Board. He is a passionate advocate for the intersection of technology, security, and patient care.

Chapter 1

The Rise of the Intelligent Health Consumer

Tom Lawry

> As in the earlier industrial revolutions, the main effects of the information revolution on the next society still lie ahead.
>
> **–Peter Drucker**

The world is becoming more intelligent and mobile thanks to the cloud, the Internet, and a cornucopia of artificial intelligence (AI)-enabled smart devices and apps.

Today four billion people are connected to the Internet, and nearly all of them do so using mobile devices (92.6%). Every day, 85% of users (3.4 billion people) connect to the Internet and spend, on average, six-and-a-half hours online.[1]

Consumers are spending more time with less effort using an exponentially expanding range of AI-enabled apps to manage and enrich virtually all activities of daily living. In the time it takes to read this chapter, one hundred thousand consumers will order an Uber through its AI-enabled app on their smartphone. To service each customer request, Uber will use AI to instantaneously predict rider demand. They will use AI to determine "surge pricing," calculate ETAs for each ride, compute optimal pickup locations, and scan for credit card fraud.[2]

As riders are taken to their destinations (likely in a Prius), they will fill time using AI-driven apps on their smartphones. They will check a travel site that predicts which day will be the cheapest to buy tickets for an upcoming

DOI: 10.4324/9781032690315-1

vacation. They will Yelp to evaluate a company before giving it their business. They'll pre-order dinner from their favorite restaurant app that makes intelligent suggestions based on their understanding of each family member's culinary likes and dislikes.

When they get home, they will use smart devices and conversational AI to provide contextual interactions between the digital and physical worlds (Hey Cortana, remind me to talk to the product team lead tomorrow when I get to the office), direct requests for physical products and services (Alexa, order more laundry detergent), and drive social interactions (Hey Siri, call my brother with FaceTime).

Despite the cost reduction and benefits of AI to the activities of daily living, when it comes to healthcare these same consumers are paying a growing portion of their healthcare costs. Even if they're fortunate to be part of an employer-sponsored plan, they'll shell out an average of $6,000 annually for a family's health insurance plan. That's just their share of the upfront premiums and doesn't include co-payments, deductibles, and other cost-sharing fees once they need care.[3]

Beyond costs, intelligent consumers are hard-wired into the convenience of online banking, shopping, and an endless supply of free or low-cost business and personal apps. All of these allow them to make things happen on their terms. Their attitudes about healthcare choices and how much administrative complexity they want to endure are altered toward having the same expectations of doing things "smartly" on their terms.

Let's look at millennials for a moment. Those born between 1981 and 1996 are now the second largest generation among commercially insured Americans. And they're on track to become the largest generation in the near future.[4] Unlike older generations, millennials walk to the beat of a very different drum. Only 68% of millennials have a primary care physician, compared to 91% of Gen Xers.[5] They are less influenced by traditional marketing but are highly influenced by social media. They are twice as likely as other generations to take actions based on health advice via social media channels or online.[6]

Another key characteristic of the intelligent consumer is that self-service or on-demand customer service is not only accepted, it's expected. A large and growing portion of health consumers want, and expect, a quick answer or resolution to a question or problem. And unlike older generations, they don't want to make a phone call to get it.

The proliferation of AI is already ever-present in the average consumer's life and is transforming how they incorporate information, interfaces, and exchanges. They are driven by the desire to not only simplify their lives, but also add flexibility and personalization into their increasingly demanding lives.

It is against this backdrop of desired efficiency, coupled with high deductible health plans requiring pre-deductible out-of-pocket spending that you should ask the question of what experiences your organization will offer these empowered consumers when they have a health-related need for themselves or their families.

The Rise of Intelligent Health Systems

As the "connected health consumer" becomes the new norm, traditional health organizations will compete with new market entrants for mind share and market share of this important population. These market dynamics will lead to the emergence of Intelligent Health Systems.

So, what exactly is an Intelligent Health System? Simply put, an Intelligent Health System is an entity that leverages data and AI to create strategic advantages through the efficient provision of health and medical services across all touchpoints, experiences, and channels.

Intelligent Health Systems will take new approaches to overcoming the age-old challenges of improving access, quality, effectiveness, and costs of health services. They will do this by being faster and smarter than similar organizations in making use of AI-enabling technologies, ubiquitous connectivity, and smart devices and systems.

From global NGOs and regional health systems to individual departments within a hospital or government-sponsored social service organization, Intelligent Health Systems will come in all sizes and forms. The defining characteristic will not be size but rather how they use data and AI to drive measurable change and improved outcomes at scale.

Other characteristics that differentiate Intelligent Health Systems from traditional health organizations include

- Blurring or eliminating historical care and service delivery boundaries by utilizing smart technologies like Internet of Things (IOT), remote monitoring and wearables, virtual visits, virtual clinical assistants, and digital twins.
- Eliminating the traditional "partitions" between health and wellness services and medical interventions with the use of predictive capabilities to dramatically increase the proactive nature of monitoring and managing the health of individuals as well as the health of populations.

- New processes and workflow models supported by AI-driven automation reduce the complexity of the medical bureaucracy that exists today to improve the consumer and patient experience.
- More effective and seamless provision of health and clinical services that are enabled through the automatic exchange of information across previously siloed data systems.
- Greater satisfaction and effectiveness of clinicians through the use of assistive intelligence that reduces administrative burden while increasing the real-time, predictive capabilities of structured and unstructured data from multiple sources to support the provision of evidence-based, patient-centered care.
- Intelligent forward-looking population analyzes that guide future development and agile adjustments to planning and service delivery to meet the needs of health consumers and patients more efficiently and accurately.
- Turning smart consumer devices into powerful diagnostic, health, and engagement tools by delivering seamless two-way connectivity to information and services considered useful and convenient to users.

As you consider your next moves on the path to becoming an Intelligent Health System here are five things to consider.

Digital Transformation Is the Onramp for Intelligent Health

Health organizations of all sizes today are investing in various digital transformation efforts. Many see this move as market strategy and differentiator. And while such investments are critical to future success, it's important to recognize that digital saturation is the new norm amongst consumers. Today, every organization is investing in digital and AI technologies. As such, your organization's digital transformation effort is less of a strategy and more a "price of admission" to compete in the intelligent health arena.

Healthcare is in a unique place in the "post-digital" world. The industry is recognizing that digital must become part of everything it does. While investments in social, mobile, analytics, and cloud (SMAC) technologies are progressing and demonstrating value, the health industry has not come as far in adoption maturity as other industries.

Today, innovative health leaders are looking at how their digital technology investments will power changes to both business models and service delivery in keeping with the new norms or the market. The future will be about full adoption of SMAC and embracing AI technologies to transform outcomes and ultimately change lives. Planning today for the post-digital world is critical as healthcare enterprises continue their digital transformation journeys.

Emerging Intelligent Health Systems are changing the game and bypassing traditional health market strategies by making better use of data and a growing array of AI-powered tools and apps to redefine how services are delivered. They leverage these assets as tools to better understand their customers as patients with a new depth of granularity and develop smartly efficient methods of providing health services to consumers on their terms.

Walgreens is a great example. They are pioneering new healthcare delivery models by leveraging cloud-enabled AI within their massive brick-and-mortar footprint (9,800 locations) and by providing a growing array of smart applications.

Their approach includes new in-store "digital health corners," telemedicine kiosks, and a digital marketplace called Find Care Now which is a desktop and mobile app that allows consumers to find and access local and digital health services including neighborhood clinics and physician house calls. The Walgreens mobile app alone has been downloaded more than 50 million times and has 5 million active users a month.[7]

Even Walmart, the world's largest retailer, is getting into the business of health by combining its retail footprint with intelligent online tools to offer a growing array of primary and specialty services that were typically the purview of traditional provider organizations.[8]

If you are working in a traditional provider or payor organization, one of the top challenges is not whether your organization will become an intelligent health system but rather how you will keep up with the demands and expectations of those you seek to serve. If you don't ... who will?

There Are No Swim Lanes in the Blue Ocean

The intelligent marketplace is not only changing consumer habits, but also erasing historical industry boundaries and encouraging new players to enter the health market. The time when the healthcare could be delineated into categories like providers, payors and retail health is giving way to a new market. It's being driven by those who seize opportunities to use their AI and data prowess to identify and curate new types of relationships with health consumers.

In the book *Blue Ocean Strategy*, authors W. Chan Kim and Renee Mauborgne, present an analytical framework that fosters an organization's ability to systematically create and capture "blue oceans"—unexplored new market areas. It's an approach used by new entrants to the health market that focuses on solutions that are both exponentially more valuable to end-users *while lowering costs*. The end goal for disruptors is to create new demand and capture uncontested market space in a way that makes the historical market players less relevant. The success of Netflix, Wikipedia, and Airbnb are all examples of blue ocean strategy.[9]

Take Amazon for example. They've set their sights on healthcare by creating a partnership with Berkshire Hathaway (an insurance and holding company) and JP Morgan Chase (a global financial services firm). Known as Haven, the stated goal of this new entity is to pool resources and expertise to explore ways to disrupt the current health market.[10]

On its own, Amazon announced the creation of Amazon Care. This pilot is initially focused on Amazon employees. They will have concierge-level access to virtual and in-person care, an intelligent telemedicine app as well as follow-up visits and prescription drug delivery to a person's home or office.[11]

Not to be outdone, Walmart jumped into the health world with a bundled set of services for Sam Club members. This includes unlimited telehealth at one dollar per visit. This AI-driven service has patients first telling their symptoms to a chatbot or automated assistant. The information gathered then gets passed along to a clinician for diagnosis and treatment.[12]

Another example is pharma giant CVS acquiring healthcare insurer Aetna in a bid to expand its footprint into transforming healthcare delivery.

The Consumerization of Health

Today a growing number of health consumers and patients are transforming from passive recipients to active participants in taking charge of their healthcare delivery. They are doing this by using online review sites and social media to choose which doctor to see, skipping hospital and traditional physician office visits in favor of health clinics built into retail outlets or company-sponsored virtual visits. For example, according to a Rock Health study, 42% of Millennials have used synchronous video telemedicine, compared to just a quarter of Gen Xers and under 5% of Baby Boomers.[13]

Look closely and you see that much of this movement is being powered by smart apps and intelligent consumer engagement activities delivered by organizations that have adopted a Blue Ocean Strategy. This revamping of the patient experience is an early example of the market moving to reward those choosing to become an Intelligent Health System.

AI Is the New UI

In the world of software and digital solutions the term User Interface, or UI, describes the way or manner in which a user engages or interacts with a system. Think about apps on your phone or computer that do this well. Now think about apps and websites that increase your blood pressure as you try and get something done with a UI that is neither intuitive, convenient, nor personalized. With this concept in mind, consider how easy or difficult your organizational interfaces make it for today's consumers to utilize your services.

It's a useful exercise to apply the UI concept to how patients and consumers interact with your organization today. How might the application of AI plus intelligent processes and systems be used to improve your organization's "UI" for patients, clinicians, and staff?

In the world today, products, services, and surroundings are increasingly customized with data and AI to cater to the individual. The challenge and opportunity in healthcare is to reimagine the existing processes and touchpoints we run consumers, clinicians, and staff through every day.

Think about this the next time you're in the grocery store weighing the option of standing in line to have a human ring you up versus using the self-checkout kiosks. Even these options are rapidly becoming obsolete as companies like Amazon apply AI to create a better experience. *Amazon Go* is an intelligent system being piloted in Seattle which allows shoppers to download an app and then simply walk out of the store with the items in their cart or basket. Instead of making the checkout process more efficient they are using AI to completely eliminate it.[14] It's also interesting to note that Amazon now owns Whole Foods. Imagine the level of convenience and disruption if such technology is applied at scale.

In the hospitality world, the race is on for smart hotel rooms. Companies like Marriot are investing heavily in Intelligent rooms geared toward personalizing the experience of each guest. From smart mirrors, lighting, and showers to devices that provide customized conversational AI, the goal is to know or anticipate the needs of each guest and then auto-generate a personalized experience based on information stored in an app.[15]

Just as Amazon is using AI to eliminate the checkout process, what might health organizations do to rethink registration, admitting, and discharge processes? Whether an exam room in a clinic, or an inpatient environment, can we take a lesson from hotel chains like Marriot in using AI to create a more personalized experience with the physical surroundings in which we see patients?

Employee Experiences Will Be as Important as the Customer Experiences

These same dynamics of a more personalized experience for health consumers and patients also extend to clinicians, caregivers, and all staff who will have an array of smart machines and knowledge systems that free them of tedious tasks and allow them to focus on practicing to the tops of their licenses.

As a new class of intelligent workers emerges those that know how to leverage smart systems and workflows will be in high demand and likely short supply. Critical to market success is recognizing that what's happening on the inside of the organization is going to be seen and felt on the outside by health consumers. Creating and maintaining a great employee experience not only supports an improved experience for the intelligent health consumer, but it also reduces the chance of other health organizations stealing your best people.

Intelligent Health Is a New Journey

While many good things have come from the traditional ways in which health and medical services are provided today, the systems we have created are struggling to keep up with the needs and demands of a changing market. These systems are being overwhelmed by a set of complex challenges which produce inconsistent quality, unacceptable levels of harm, too much waste, and unsustainable costs.

Increasingly, many health leaders have a vision and aspirations for being better. And while everyone is early in their journey toward Intelligent Health Systems the market is beginning to take hold as innovative leaders leverage smart technologies with the goal of having every contact with consumers and interface with staff be simpler, smarter, and more efficient.

Just as having a website and using social media doesn't make your organization an Internet company, the use of AI and smart applications doesn't make an organization an Intelligent Health System. Becoming an Intelligent Health System sets a high bar for health leaders to clear when it comes to developing or changing core organizational competencies involving technology, data, people, and delivery mechanisms. Such systems will be built by those who are prepared to think differently about how AI will drive changes to clinical and operational processes along with a laser focus on improving the experience for consumers, clinicians, and staff.

Sidebar: The DNA of Intelligent Health Systems – An Interview with Harry P. Pappas, CEO, Intelligent Health Association

When it comes to seeing and understanding the rise of Intelligent Health, there are few people in the world with as much history and experience in its evolution as Harry P. Pappas, Founder and CEO of the Intelligent Health Association (IHA). As a global, technology-centric organization, the mission of IHA is to help drive the adoption and implementation of intelligent technologies throughout the health ecosystem. IHA does this through advisory and educational services to clinical and health leaders in the United States, Europe, and the Middle East.

> Smart technologies are already having a huge impact on consumer health and wellness habits," says Pappas. "As the complexities of healthcare delivery increase and the sophistication of health consumers in using smart technologies grows, there's an imperative for health organizations to quickly move towards the adoption of AI and intelligent solutions to keep pace. In some ways the market is getting out ahead of many traditional providers which creates both a threat and opportunity going forward.

With topics ranging from AI, virtual reality, 3D printing, and intelligent voice technologies, IHA is routinely interacting with clinical and health leaders to forge collaborative efforts that champion the incubation and adoption of intelligent solutions that improve the provision of health and medical services across the spectrum of care settings.

> In our experience, the single best predictor of the success of any intelligent health initiative is the level of understanding and participation of clinical and C-suite leaders. It's critically important that leaders get their minds around how intelligent health solutions change how work is done, how things are managed and how relationships with consumers are built and solidified.

When it comes to current uses of AI, IHA is seeing interest in, and working with, health organizations in the following areas:

- Internet of Health Things (IoHT): Use of intelligent wearable medical technology that helps patients and clinicians monitor vital signs and symptoms with machine learning built in to assess and predict variables important that help proactively manage health and intervention activities.

- Infrastructure optimization: While optimizing the management of expensive facilities is not as sexy as clinical use cases it's an area driving measurable value. Use cases range from the use of technologies to create smart buildings to elevators that learn usage patterns that are then used to improve throughput efficiencies.
- Conversational AI: The advent of chatbots and intelligence voice devices is gaining traction in the health arena. In the home, they are increasingly being used as a way of engaging consumers in managing their health. Use cases include real-time information and support for managing weight and stress.

Notes

1 https://www.nielsen.com/us/en/insights/reports/2018/connected-commerce-connectivity-is-enabling-lifestyle-evolution.html

2 Mansoor Iqbal, Uber revenue and usage statistics 2019, http://www.businessofapps.com/data/uber-statistics/#1

3 John Tozzi, The cost of health insurance for a family hits a record, passing $20,000 a year, Los Angeles Times, 2019, https://www.latimes.com/business/story/2019-09-27/health-insurance-costs-surpass-20-000-per-year-hitting-a-record

4 Millennial health trends found in report concern doctors, Pittsburgh Business Times, 2019, https://www.bizjournals.com/pittsburgh/news/2019/10/08/millennial-health-trends-found-in-report-concern.html

5 Ibid.

6 Bill Coontz, The millennial mandate: How 'Generation Y' Behavior is shaping healthcare marketing strategy, MediaPost, https://www.mediapost.com/publications/article/253264/the-millennial-mandate-how-generation-y-behavio.html

7 Heather Landi, Walgreens' CMO on the retailer's expanding healthcare role, Microsoft partnership, FierceHealth, 2019, https://www.fiercehealthcare.com/tech/walgreens-cmo-talks-digital-healthcare-and-microsoft-partnership?mkt_tok=eyJpIjoiWlRSaE9XTTNaREUwT0RabCIsInQiOiJPVmhlSm9PS2NFTWRFRTYzV2gxWVRZMUV1RlpQS3Fiem84VHpwaXN1N213U1h5UU5MY281-azZCVkh0M1A2cXJKNDRDcEtGWm1hNkI3Z3RHakdmOUorbFcxNnQ4MUh1bnJZT2JZXC81N3lcL3-dIVWtTMW4zdHlkb2c0S1NiRmNEVGtsIn0%3D&mrkid=1009087

8 Christina Farr, Walmart tests dentistry and mental care as it moves deeper into primary health, CNBC, 2019, https://www.cnbc.com/2019/08/29/walmart-is-piloting-health-clinic-at-walmart-health-in-georgia.html

9 W. Chan Kim, Renée Mauborgne, Blue Ocean Strategy, Harvard Business School Publishing, 2015

10 Angelica LeVito, Amazon's joint healthcare venture finally has a name: Haven, CNBC, 2019 https://www.cnbc.com/2019/03/06/amazon-jp-morgan-berkshire-hathaway-health-care-venture-named-haven.html

11 Darrell Etherington, Amazon launches Amazon Care, a virtual and in-person healthcare offering for employees, TechCrunch.com, 2019, https://techcrunch.com/2019/09/24/amazon-care-healthcare-service/

12 Anne D'Innocenzio and Tom Murphy, Walmart's Sam's club launches health care pilot to members, AP Business Writers, 2019, https://www.usnews.com/news/us/articles/2019-09-26/walmarts-sams-club-launches-health-care-pilot-to-members
13 Millennials are leading the way in digital health, Huffington Post, 2017, https://www.huffpost.com/entry/millennials-are-leading-the-way-in-the-adoption-of_b_59f4bd62e4b05f0ade1b57a9
14 Nick Wingfield, Inside Amazon Go, a store of the future, New York Times, 2018, https://www.nytimes.com/2018/01/21/technology/inside-amazon-go-a-store-of-the-future.html
15 Gary Diedrichs, The "Smart" hotel room race is on, Smart Meetings, 2017, https://www.smartmeetings.com/technology/104502/smart-hotel-room-race

Chapter 2

Fundamentals of the Intelligent Health System

Paul H. Frisch

Introduction

The rise of the Intelligent Health System is a continuous evolution and expansion of hospital intelligence focused on establishing a standardized and seamless continuum of care which improves patient outcomes and enhances patient care, safety, and satisfaction as patients transition through the many aspects of today's healthcare process. The design of Intelligent Health System provides and manages care across all stages of the care process, from triage and emergencies, diagnostics, and specific treatment procedures, including surgeries, radiation therapy, chemotherapy, in-patient care in the hospital setting, and follow-up or post-treatment care in the ambulatory setting at regional locations or physician offices. Long-term care is possible at home via telemedicine applications and remote monitoring. The overall patient outcome and experience is directly based on the specific processes which occur during the many phases of care which take place in multiple locations and are experienced over many visits or sessions.

How the patient and their family members perceive their care process is based on the clinical outcome, quality of care, effectiveness, confidence, and satisfaction associated with their overall care processes. The evolution of the Intelligent Health System has focused on utilizing and exploiting new, evolving, and diverse technologies integrated into the clinical environment, optimizing operational and business workflows, focused to enhance and

 DOI: 10.4324/9781032690315-2

improve the patient care process. These technologies enable enhanced workflows optimizing the clinical and business operations providing a seamless data environment enabling and standardized care process regardless of the level of care or clinical area providing treatment and patient management.

This chapter highlights the diverse technologies, clincial applications, and integrations used to create an enhanced and seamless patient care, optimizing workflows, enhancing patientsafety, validating information integrity and accuracy.

Drivers for Hospital Intelligence

Over the course of the last decade, hospital-based healthcare systems have continued to experience an increase in patient acuity levels driving an increase in the number of critical care monitored beds, emergency room visits, and surgical cases [1–3]. With the advent of diminishing family practices, many patients use hospital emergency departments as their primary method of general healthcare significantly impacting patient volumes, flow, and throughput. In 2019 the Coronavirus (COVID-19 pandemic) stressed hospitals beyond their capacity with large numbers of patients presenting at Emergency Departments which directly resulted in an increasing percentage of admissions. This in turn stretched and exceeded the medical resources, including patient beds, medical equipment, supplies as well as the medical staff necessary to manage these patients effectively [4]. As a result of these developing circumstances, telemedicine and telehealth applications have driven an increased demand for home health and monitoring applications.

These factors have been identified or associated as a contributor to the increasing number of errors, including medication errors [5,6], device and systems operating errors, and errors associated with patient identification, resulting in an increased risk to the patient.

These trends are significant and have raised high levels of concern within regulatory agencies and hospital accreditation organizations, such as the Centers for Medicare and Medicaid Services (CMS) and the Joint Commission (TJC). In response, the Joint Commission has established guidelines (National Patient Safety Goals) to address these patient safety concerns [7]. Since 2010 patient safety goals have targeted developing processes to minimize these errors and optimize patient safety.

To compensate for the changing healthcare environment, hospitals found it necessary to critically evaluate and optimize their clinical and business processes to enhance patient diagnostics, treatment, care, optimizing resource utilization, and patient throughput. This has driven an increased interest and reliance on new, evolving, and enabling technologies which can be used to enhance and optimize clinical and patient care processes. The trend to exploit and integrate these diverse technologies into the healthcare environment has defined the pathway for creating today's Intelligent Health Systems.

Objectives of the Intelligent Health System

The fundamental objective of the Intelligent Healthcare system is to establish a seamless and standardized patient care environment across the entire continuum of care, including the Intelligent Hospital, Ambulatory care, and in the home. The Intelligent Healthcare System focuses on the integration, development, and deployment of a diverse combination of technologies, which includes medical devices, clinical applications, and IoT devices all integrated to seamlessly share and deliver actionable information across the continuum of care. These enabling technologies provide the tools and mechanisms to establish connectivity, unique identification, communications, association, tracking, and the delivery of information across the entire patient care environment.

Medical devices and clinical applications can share information in a variety of ways, most frequently via the hospital network and wireless infrastructure where clinical data, alarms, and other relevant information are associated with specific patients and reside in shared or within common data archives. This enables patient-specific data to be associated with care providers and clinical staff and be aggregated, processed, filtered, validated, and delivered to any location or IoT device within the environment of care. Critical data, including physiological parameters, alarms, and alerts, are available to analytic software such as decision support or AI systems and can be directly delivered to clinical staff at the point of care enabling rapid response and an enhanced level of patient care. Aggregated patient-specific information can additionally be transmitted and displayed or visualized on dashboards in many formats and at a variety of locations to enhance to provide board view of the overall status of the patient care environment to optimize and prioritize clinical workflows and business processes.

The objectives of the Intelligent Health System are outlined as follows:

- Increased Data Accuracy and Availability
  - Electronic Medical Records
  - Shared Medical Device Interconnectivity and Integration
  - Global View of the Patient/Unit
- Rapid Information Dissemination and Response
  - Physiological Data
  - Alarm Event Management
  - Advanced Communication Platforms (PDA, iPhone, etc.)
  - Location Content
- Enhanced Error Checking and Validation
  - Identification and Association
  - Location Content
  - Correct Information
- Optimization of Workflow and Resource Management

The design, development implementation of the Intelligent Health System is influenced by many factors. Few health systems have the opportunity of designing a health system from scratch, where technologies and required budgets can be developed simultaneously. Most health systems are forced to develop strategies to implement the components necessary over many years of planning and budgeting and adapt the design process to accommodate and incorporate legacy systems already in use within the healthcare system. The planning and prioritizing of the technologies represent a challenge and require a longer-term strategy and institutional commitment.

The design of an Intelligent Health System enables the development of clinical workflows and business processes that focus on establishing patient-centric or patient-specific solutions for diagnosis, treatment, monitoring, and the overall care standardized across all areas of the health system. The patient care process frequently starts either in the physician's office or in an emergency scenario such as the ambulance and or in the Emergency department where initial triage and immediate care are provided. Complex diagnostic decisions are supported with data from specialized imaging modalities and clinical laboratories. Based on the diagnosis, specific treatment plans are developed which can include surgical procedures or specialized treatments such as radiation therapy, dialysis, chemotherapy, physical therapy, etc. Post-treatment patient monitoring and recovery continues through a variety of care environments which can include the acute care/ICU, step down, ambulatory care/general care as well as monitoring of patient status post-discharge in the home.

Integrating Enabling Technologies into Patient Care Environment

The vision of Intelligent Healthcare focuses on establishing a methodology to seamlessly collect, share, and distribute accurate and validated information across the entire spectrum of the environment of care. To establish an efficient and effective seamless patient-centric environment of care, the Intelligent Healthcare System must minimize the deployment of isolated solutions and focus on developing broad-based standardized institutional solutions. These solutions are dependent on the establishing the following standardizations:

- Unique Identification
  - Patient, Staff, Supplies, etc.
  - Care Locations
    - Treatment Areas
    - Unit, Room definitions
- Patient Association Process to Devices, Staff
- IoT Devices Delivery
  - RTLS
  - Communication Device
- Infrastructure
  - Network, Wireless, Real-Time` Locations

Establishing a unique identification methodology is fundamental and enables the use of many diverse technologies to be associated with a patient and linked to a variety of archives, processing engines, dashboard, and delivery mechanisms enabling real-time data exchange via a high-speed, high bandwidth wired, and wireless infrastructure. The interconnectivity between medical devices, supporting sensors, IoT devices and technologies as well as clinical applications enables a broad spectrum of patient-centric solutions to be developed and deployed. This interconnectivity enables large volumes of data to be collected, archived, aggregated, filtered, and processed to enable critical and actionable data to be rapidly shared. The data can be delivered to a combination of locations including a central control room, dashboards, and directly to the point of care via a variety of hand-held communication devices. The table below outlines the types of devices, technologies, and applications which create the many use cases that are exploited to create a dynamic healthcare environment.

- Advanced Medical Data Systems
  - Transducers and Sensors
  - Wearables and Remote Patient Monitoring
  - Robotics
- Networked and Wireless Infrastructure
- Time Synchronization
- Patient-Centric Identification
- Centralized or Shared Data Acquisition – Connectivity of Devices
  - Physiological and Alarm Data Management
  - Intelligent Data Processing
- Visibility In Healthcare
- Location-Based Systems, RFID, RTLS, BLE
  - Device Tracking
  - Workflow Monitoring
- Infection Control
- Clinical Applications and Workflows:
  - Medication Administration
  - Sample Collection
- Video and Imaging
- Global/Institutional Overview and Visualization of Rooms/Units
- Enhanced Communications
- Point of Care Information Delivery
- Inventory Management

Components of the Intelligent Health System

Advanced Medical Systems, Devices, Transducers, and Sensors

Recent innovations have enabled extensive advancement in a broad base of medical devices, transducers, and sensors enhancing patient care across all clinical care modalities, diagnosis, treatment, care delivery, and recovery as outlined:

- Treatment Devices
  - Radiation Therapy
    - Linear Accelerators
    - Proton
    - High Dose Radiation
  - Laser Systems

- Diagnostic Systems
  - Imaging Devices
    - CT, MR, PET, etc.

- Patient Care
  - Physiological Monitoring
  - Ventilation Systems
  - Infusion Systems
- Robotics
  - Surgical Robotics
  - Disinfection/Cleaning Systems
  - Delivery Systems
  - Remote Consulting/Telemedicine
- Telehealth/Telemedicine Devices

The individual specifics of medical devices vary based on manufacturer and specific device function and use are outside of the scope of this publication. However, many Biomedical/Clinical Engineering books, publications, and manufacturer literature are available to provide comprehensive detail on specific medical device operational capability and functionality [8–10]. Please refer to the board base of literature and data available on medical devices and systems.

However, it is important to note that within today's Intelligent Health System medical devices are no longer stand-alone devices operating asynchronously. These devices are now interconnected and integrated with other medical devices, clinical applications, and IoT devices which are networked and wireless creating complex medical data systems. These systems provide decision support, data archival, and data delivery. This complex interconnectivity has defined the requirements for a high-speed infrastructure/

networking architecture, accurate time synchronization, and a standardized and unique identification to provide the required operational functionality throughout the Intelligent Health System.

Complex Medical Data Systems: Physiological Monitoring

An example of such a complex data system is the physiological data system illustrated in Figure 2.1.

The overall system architecture is based on a bedside physiological monitor operating asynchronously and capable of measuring and acquiring patient physiological parameters including heart rate, ECG, blood pressure, end-tidal CO_2, pulse oximetry, temperature, etc., and displaying the information on bedside monitor. The monitor can be networked to multiple monitoring units typically within a patient care unit via a dedicated standard network or via wirelessly infrastructure sharing data between these systems. Multiple monitors pass the clinical parameters to one or multiple central stations coupled with dashboards within the unit as necessary. All the central stations can be further networked creating a dedicated site network. This network is part of the

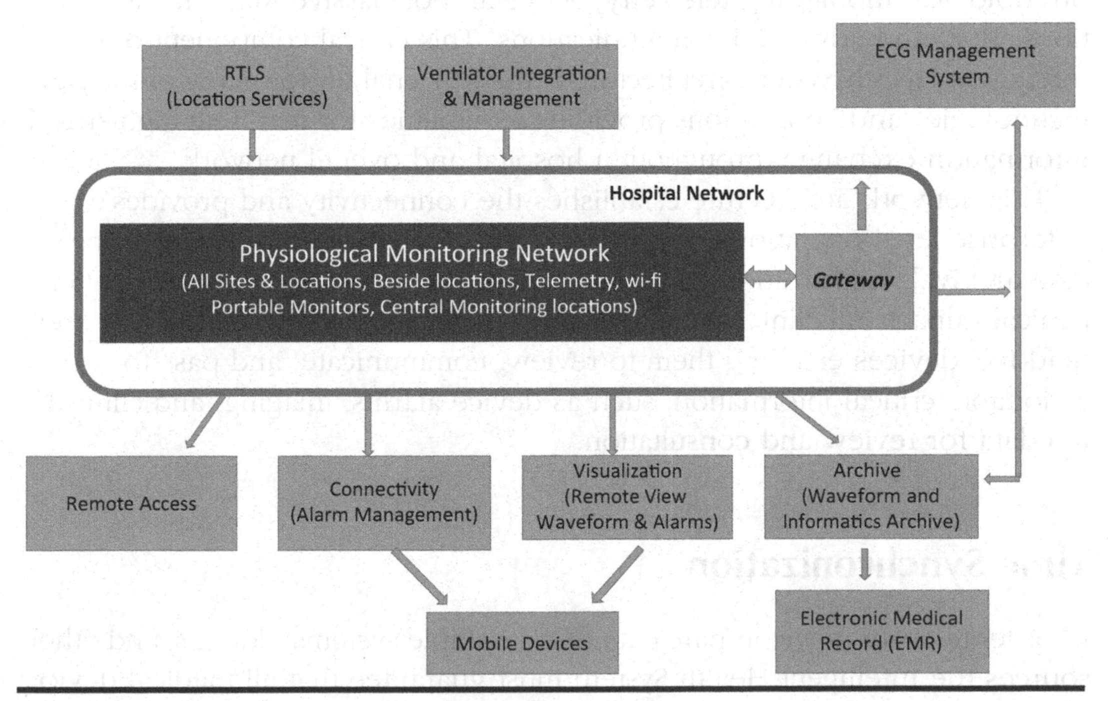

Figure 2.1 Complex integrated medical data systems.

hospital network architecture of an isolated network connected via gateways to the hospital network backbone. These configurations enable actional data to be available at other sites and to multiple clinical applications. In this case, the physiological monitoring network interfaces and passes data and alarms to remote monitoring systems, a long-term data archive, interfaces to ventilators or other devices, archives data to the patient record in the Electronic Medical Record (EMR), and enables direct communication and delivery of information to a variety of locations including the point of care via a variety of IoT devices. This complex medical data system enables data to be aggregated and seamlessly available across the institution, integrating with multiple clinical applications to provide an enhanced level of patient care [11].

Network and Wireless Infrastructure

The fundamental component to provide a seamless shared data environment is the underlying infrastructure providing the high-performance network consisting of a seamless network architecture including a blend of multi-carrier cellular, Wi-Fi services, and dedicated wireless frequencies. Together these support the specific functionality, required for medical devices, such as physiological monitoring telemetry, active and or passive RFID, RTLS solutions, BLE, and advanced communications. This critical component of the Intelligent Health System architecture is the key enabler for many of the new technologies and applications providing capabilities for real-time high-speed information exchange throughout a hospital and overall network.

This network architecture establishes the connectivity and provides enterprise-level operational intelligence and the opportunity for real-time data archival, visualization, and clinical response enabling a positive patient clinical impact. All clinicians and medical staff can now utilize mobile handheld IoT devices enabling them to review, communicate, and pass forward actionable critical information, such as device alarms, imaging, and clinical lab data for review and consultation.

Time Synchronization

In order to accurately compare data from multiple systems, devices, and other sources the Intelligent Health System must guarantee that all medical device data parameters and information are time synchronized throughout the entire care environment. This ensures that treatment and care decisions are made effectively based on accurate patient data and clinical documentation. Time synchronization is an effective way to ensure operational efficiency within the healthcare environment and a standard to ensuring accurate, synchronized time throughout the healthcare network ensuring that treatments and medications are administered at the appropriate scheduled times.

Mobile IoT Devices

The network and wireless infrastructure not only supports the integration of medical devices and clinical applications but also enables data to be relayed to commercial products such as mobile hand-held and wearable devices, provide the ability to readily communicate, receive, and deliver critical information, including alarms, clinical parameters, or lab data. These hand-held devices are commonly used as point-of-care delivery devices but can simultaneously integrated into the hospital phone system enabling incoming and outgoing calls and the ability to support real-time remote consultation.

Patient-Centric Identification and Association

The identification and association of patients with medical devices, clinical data, care providers, pharmaceuticals, etc. is a vital component to ensure patient safety, the validation of clinical processes, and optimizing operational workflows [12]. This process is required by many clinical and business applications, including medication administration, and sample collection and is required for establishing connectivity of most medical devices identifying patient parameters and alarms to a specific patient. Operationally, this process is performed and repeated many times for each patient, requiring valuable time and introducing a potential for inconsistencies and errors. Fundamental to creating a seamless data environment is a standardized and highly accurate methodology to identify and associate patient-specific data with informational flow. This association methodology defines and establishes the dynamic relationship between staff, patients, clinical systems, medical devices, pharmaceuticals, and supplies used in the patient's treatment and care. Patient-centric identification, as illustrated in Figure 2.2, can rely on the use of multiple technologies, including barcodes, radiofrequency identification (RFID, both active and passive), infrared, and BLE to support the bedside association

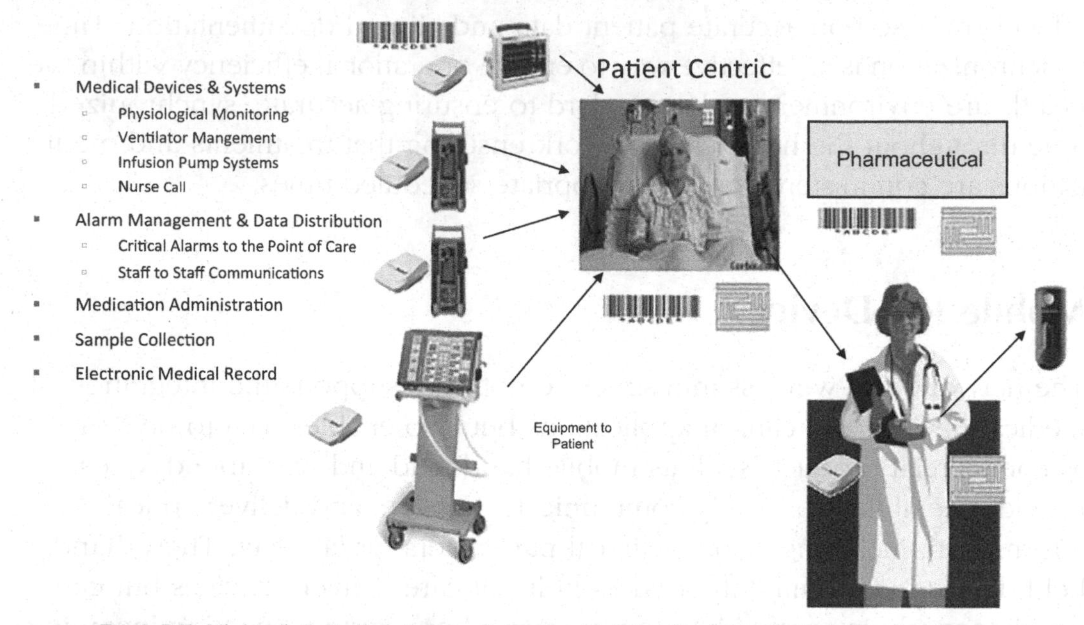

Figure 2.2 Dynamic patient-centric identification and association.

process. As previously outlined, it is important to note that standardization and naming of physical space/locations and devices is also vital to enable precise location identification and tracking.

One approach enables the association process at central locations such as nursing stations or staff work areas. The association application integrates with the hospital's ADT systems for patient information, active directory for staff information, and the biomedical device/asset database for equipment-related information. This enables associations to be defined centrally. However, many of the critical identification processes, such as connectivity, medication administration, sample collection, and imaging optimally, occur at the bedside or point of care, which ideally uses identification technologies (bar-codes, RFID, BLE, etc.) directly at the point of care. This bedside association can be stored and distributed to the appropriate applications or medical systems to minimize redundancy.

Connectivity, Interoperability, and Integration

As the acuity of patients increases, patients are typically connected to an increasing number and variety of medical devices including ventilators, infusion pumps, cardiac output devices, etc. It is vital that data from these discrete devices and systems are time synchronized and patient-specific data be aggregated to provide a comprehensive view and interpretation of the patient status and condition. The process of acquiring clinical or physiological parameters and alarm information from multiple medical devices and passing this data to a central processing system or archive defines the connectivity application. The processing engine or connectivity solution retains the association relationships and stores, manages, and delivers the patient-specific information as a function of the association.

Advanced Communications and Point-of-Care Clinical Data Delivery

The Intelligent Health System is also highly reliant on advanced communications methodologies and location technologies, as illustrated in Figure 2.3, to maintain rapid data dissemination and capabilities for clinical consultation

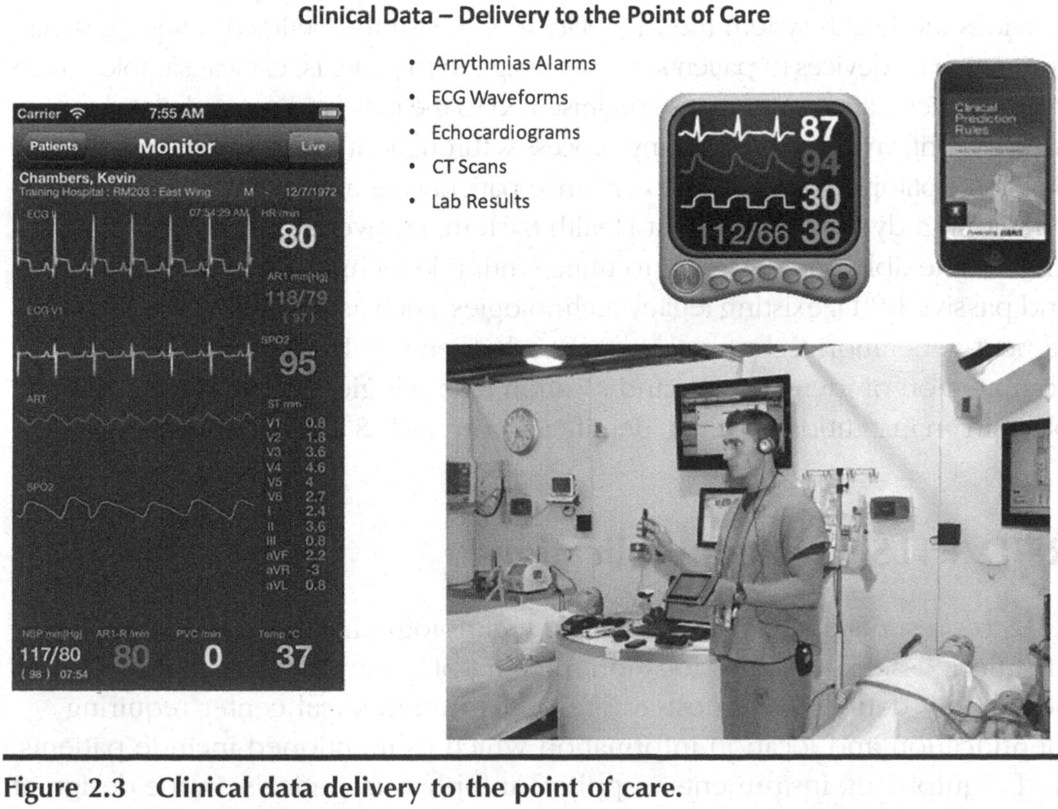

Clinical Data – Delivery to the Point of Care

- Arrythmias Alarms
- ECG Waveforms
- Echocardiograms
- CT Scans
- Lab Results

Figure 2.3 Clinical data delivery to the point of care.

and decision support. As the acuity of patients increases, both physician and nursing time at the bedside needs to increase proportionally. Therefore, the timely delivery of patient information, such as physiological data, alarms, laboratory results, or imagining, is key to optimizing patient treatment and care. With the deployment of a robust network infrastructure, connectivity, and time synchronization coupled with patient-centric identification the healthcare facility is ready to implement and deploy point-of-care delivery solutions. The use of commercial hand-held devices, such as iPads, wireless phones, and tablets, operating on the hospital's wireless infrastructure, enables a cost-effective method for staff to receive, review, and request clarification on the information directly from the point of care.

Real-Time Location Systems

Technologies such as radiofrequency identification (RFID), real-time location systems (RTLS), infrared (IR), ultrasound and low energy Bluetooth, etc., are commonly used to provide the unique identifier for institutional assets, which can include devices, supplies, patients, staff, and pharmaceuticals. The unique identification provides the health system the capability to establish the outlined unique associations, such as devices to patients, nurses assigned to patients, clinical samples taken from a patient, and medications administrated to the patient. These relationships are used to confirm and validate many processes throughout the healthcare system and develop optimized workflows to enhance patient care and safety.

Within a dynamic Intelligent Health System, an overall institutional design must be flexible and designed to utilize multiple technologies including active and passive RFID, existing legacy technologies, such as barcodes and IR, as well as next-generation technologies, such as low energy Bluetooth (BLE). This combination of identification and location technologies will all have a shared place in an institution's overall identification and RTLS as illustrated in Figure 2.4

RFID/RTLS Enabled Applications

RTLS represents a significant enabling technology that permits unique identification, association, and location tracking of institutional assets. Assets can be broadly defined as almost any resource in a medical center requiring identification and location information which as mentioned include patients, staff, equipment, instruments, supplies, and pharmaceuticals. In the design of

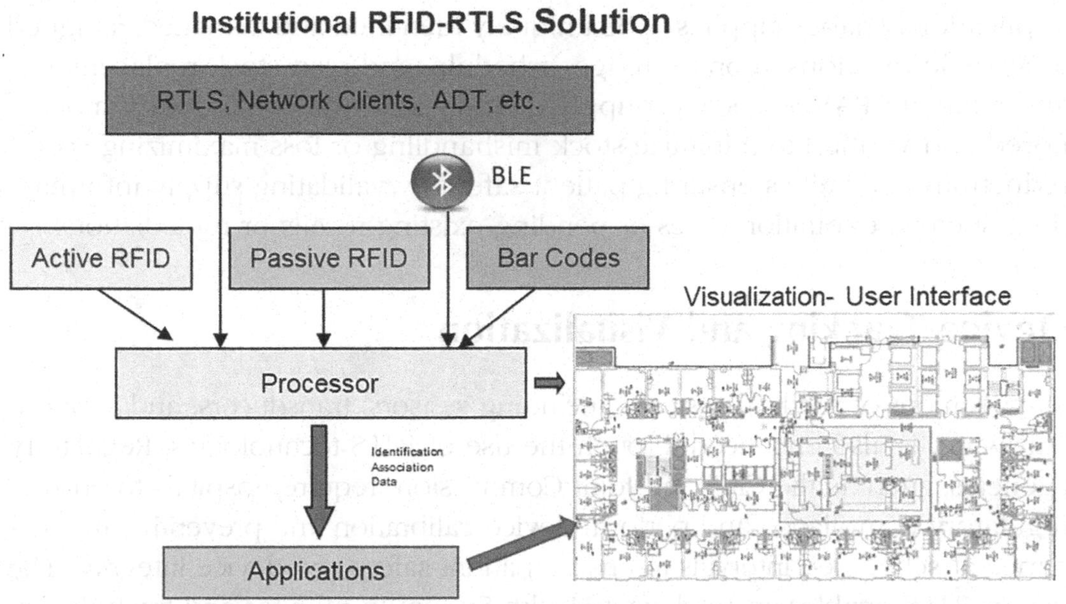

Figure 2.4 Design architecture of an institutional location system.

an Intelligent Health System, coupling identification and location information with existing processes and applications can help characterize and optimize the workflow processes used in healthcare delivery. Users of RFID/RTLS solutions quickly realize that these technologies can be applied to hundreds of use cases and applications which have the potential of improving and optimizing a broad base of processes and operations.

Within the overall Intelligent Health System environment, RTLS applications can be classified into several primary broad categories:

- Supply Chain Inventory Management
- Medical Device Tracking and Visualization
- Utilization
- Workflow Monitoring and Analytics
- Process Validation
- Regulatory Compliance

Supply Chain, Inventory Control, and Management

Inventory management and supply chain is perhaps one of the most obvious applications for identification and location-based technologies. Supply chain applications enable supplies to be uniquely identified and managed, assigned to specific locations, monitor usage, schedule, and plan stock replacement, and maintain PAR levels, etc. Supply quantities can be continuously monitored and verified to minimize stock mishandling or loss maximizing cost reductions, as well as ensuring patient safety by validating supply information, such as expiration dates or pending existing recalls or remediations.

Device Tracking and Visualization

Management of medical systems, including sensors, transducers, and support hardware, is also enhanced through the use of RTLS technologies. Regulatory agencies, such as the CMS and Joint Commission, require hospitals to ensure regulatory compliance and perform device calibration and preventive maintenance at scheduled intervals to ensure patient safety and device integrity. The use of RTLS enables an Intelligent Health System to monitor and track device location ensuring all maintenance is performed as necessary and in a timely manner based on manufacturer requirements and schedules. Active tags are commonly associated with critical highly mobile devices such as infusion systems providing global institutional visibility and real-time tracking.

However, many devices are small and not conducive for large active tags such as smaller device modules, sensors, transducers, or probes. In this case, smaller form-fitting tags are advantageous for these applications, as illustrated in Figure 2.5. Hand-held readers or dedicated antenna systems need to be used to identify and locate these smaller devices. Ideally, the information is merged to provide a global view of all of the institution's assets and inventory. This use case specifically highlights one of the examples of why an institutional solution needs to support multiple location-based technologies as part of the overall RTLS solution.

Asset Visualization

The use of RTLS enables medical devices and or other assets to be dynamically tracked providing a view of the institution assets globally and within specific patient care areas down to the room level. To optimize rapid response ideally, devices are identified based on their availability for use. Coupling device status information with the RTLS system location provides a significant value to search and retrieve only devices not in use. The optimum design approach focuses on developing a direct integration with medical

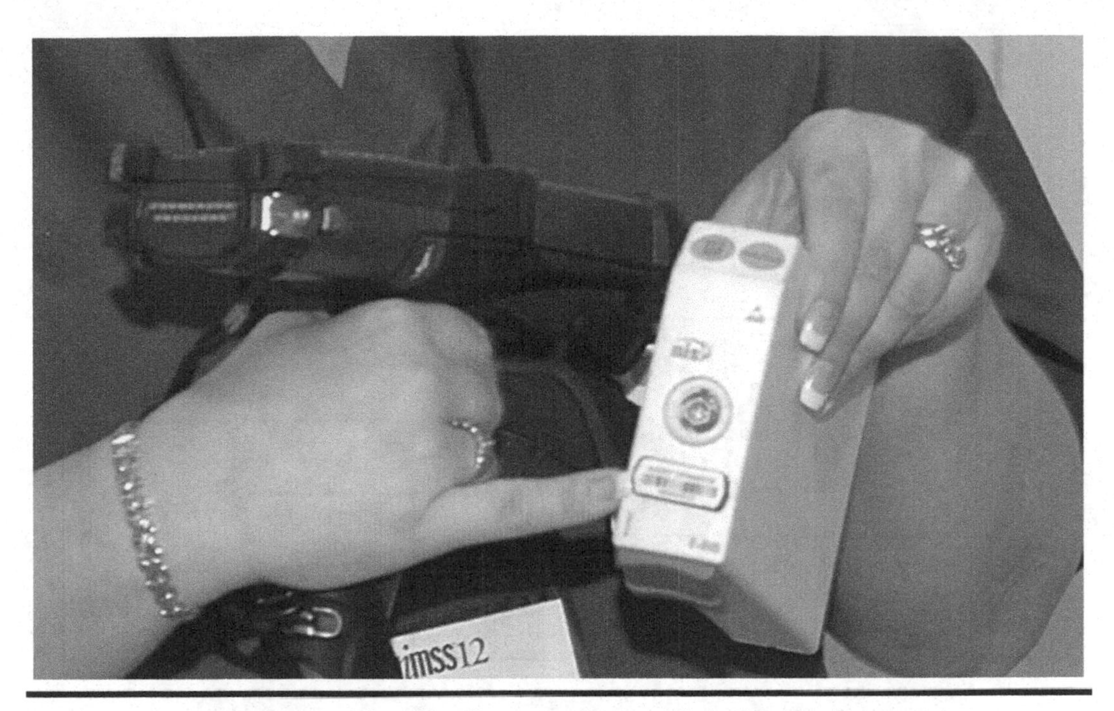

Figure 2.5 Passive RFID identification tags for small devices and sensors.

devices to extract various operating parameters, such as operating status, clinical parameters, or alarm information. This is typically a complex development effort and is difficult to perform across a broad base of medical devices or systems. However, for specific medical devices which are in high demand and commonly associated with frequent searches such as infusion systems, this can have significant value.

An alternate approach applicable to all medical devices is utilizing the event button commonly available on many RFID tags. The event buttons can be used to set the specific state of a device and can be defined to represent a variety of functionality or status (in use/available). This method does present some limitations since it requires supporting workflows to ensure the device state or status is accurately set. The accuracy of the information becomes highly dependent on the compliance of the staff to set or use the event button correctly; however, this solution is readily available and can be applied to a broad base of devices or other assets without additional cost or any significant development effort. Devices could now be visualized via status identifiers such as a color-coded scheme on the floor map visualization, red as unavailable and green as available. This approach provides a robust and accurate representation and view of device status as illustrated in Figure 2.6.

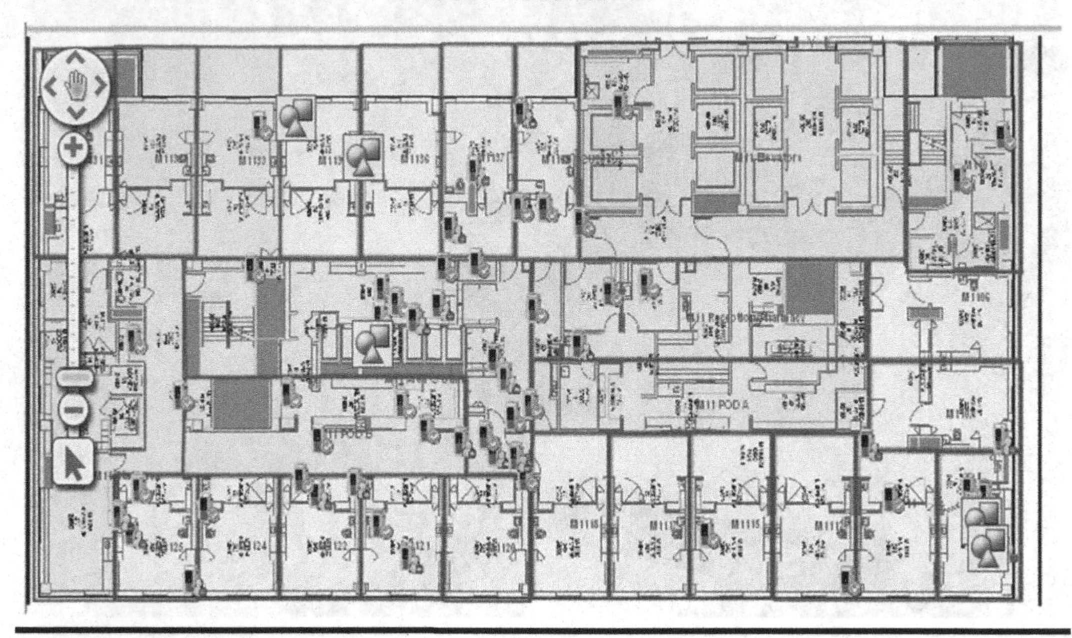

Figure 2.6 Asset visualization.

Utilization Statistics

Once device status and availability as a function of time are identified for each infusion system or medical device, it is possible to compute the overall utilization of the device. The information can be trended by overall institution, specific location, or patient care area. This can be useful to administration to make strategic purchasing, device distribution, or allocation decisions.

Process Verification and Validation

One of the most critical operational capabilities of location-based solutions is to provide a mechanism to enable validation and quantification of critical processes. Location technologies such as RFID/RTLS provide a tool or mechanism to enable validation or confirm that specific processes have occurred minimizing errors and enhancing patient safety as outlined in the following table:

■ Inventory Validation	Ensure medical device or asset inventory accuracy
■ Hand hygiene compliance	Monitor and ensure hand hygiene compliance
■ Surgical instrument confirmation	Ensure all surgical instruments have been accounted for and recovered minimizing patient incidents due to instrument retention
■ Surgical sponge verification	Ensure all surgical sponges are accounted for minimizing surgical incidents due to sponge retention
■ Verification of cleaning processes	Ensure all devices undergo a cleaning process prior to patient use
■ Medication administration	Confirm the 5 patient rights prior to medication administration
■ Sample collection	Ensure all samples are associated with the correct patient
■ Process validation	Ensure cleaning and sterilization procedures and processes

Workflow Metrics and Optimization

A valuable and important metric is the ability to quantify workflow processes and define the corresponding impact on patient care and safety establishing a pathway and measure for continuous process improvement. Tracking patients and staff in real time and correlating their location with specific events provides many options for workflow monitoring, management, reporting, and optimization. One specific metric of interest is the response times of staff to events, including patient calls, codes, critical alarms, etc. Availability of this type of data provides a quantitative metric on the level of staff-to-patient interaction which can be directly correlated or representative of patient care quality and consequently an indicator of potential patient satisfaction. This information can also be correlated to staffing requirements to define and maintain an institution's optimal or desired level of patient care and interaction.

As illustrated in Figure 2.7, based on a study performed at Memorial Sloan Kettering Cancer Center [13] to study the feasibility of using RTLS to monitor patient flow within an Ambulatory setting. Specific areas were outfitted with localization hardware to enable a choke point architecture defining specific locations such as registration, waiting area, treatment rooms, etc. [9]. The ability to quantify workflow processes through these instrumented locations generated specific information on the status or impact on patient care and

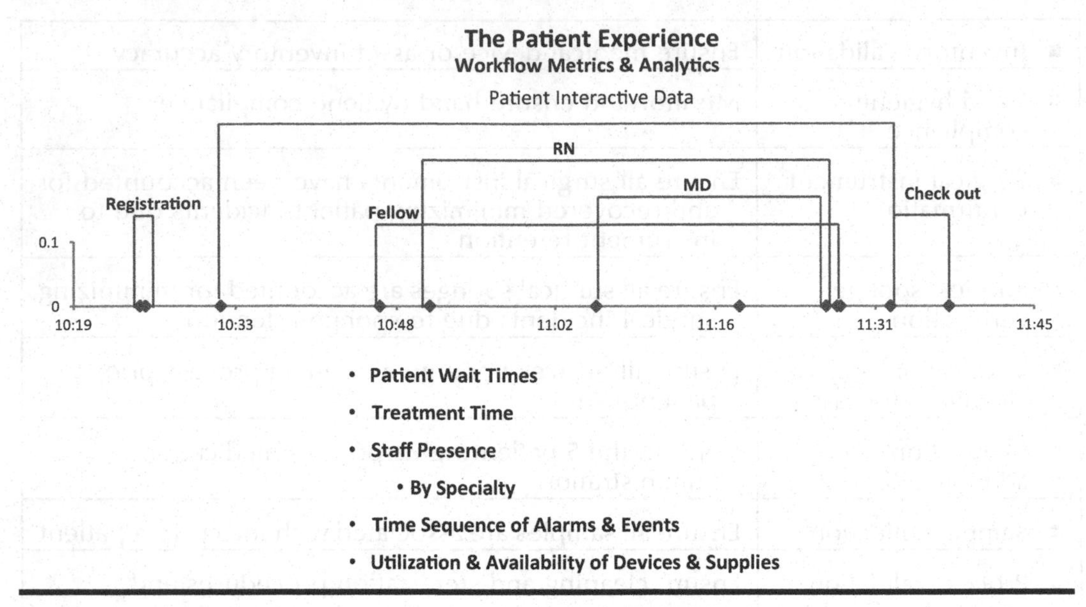

Figure 2.7 The patient interactive data and experience.

safety. This information provides valuable data and establishes a pathway for process enhancement and improvement of the patient care workflow and patient satisfaction.

Technology Management and Support in the Intelligent Health System

Development of an Intelligent Health System requires a clear vision and strategic plan which clearly outlines the objectives and defines required technologies, integrations, deliverables, and data flows. Organizations that are fortunate enough to be designing and building a new healthcare facility from the ground up can design and implement a facility based on an institutional approach standardizing medical devices, enabling technologies, and the networked and wireless infrastructure throughout. This standardization enables technology management support and maintenance solutions to be established and defined during the design process and can frequently be built in as part of the over design and deployment process.

Most organizations do not have this opportunity and require the design, development, and implementation of a fully integrated and seamless healthcare institution as ongoing transition plan of how to incorporate new, evolving, and enabling technologies into the various treatment modalities and facilities which includes integrating with existing legacy systems and workflows. To ensure a successful ongoing evolution of the Intelligent Health System, there must be a strong focus on the support, maintenance, and management strategies of this highly integrated, dynamic, and complex technology-based environment. This will require the creation of new operational support and management strategies with well-defined inter-department synergies, collaborations, and responsibilities all of which need support through administration.

In today's Intelligent Hospital, these complex medical data systems utilize multiple technologies which if not properly designed, deployed, and managed can become a double-edged sword adding significant complexity to the patient care environment. These systems and technologies create dynamic and unique solutions providing an interactive seamless data flow which results in large volumes of data with potential issues and impact associated with data integrity and cybersecurity. Technology Management within Intelligent Health Systems will focus on an adaptive Biomedical Engineering program, expanding from the traditional Clinical Engineering role to include integrated electromechanical engineering, computer science/information technologies, and facilities management skill sets. New methodologies for managing complex systems, verifying functionality and performance, repairs, preventative maintenance, upgrades, and clinical staff education will need to be developed.

The following highlights some of the expanded operational issues that need to be considered as part of support and management roles and responsibilities within today's Intelligent Health System.

- Management of Big Data
  - Processing, archival, and recovery
  - Physiological Parameters
  - Laboratory Results
  - Radiological Information and Imaging
- Cybersecurity
  - Vulnerability Assessment and Reduction
  - Threat Monitoring
  - Attack Response and Recovery
- Management of Electromagnetic Environment
  - Monitoring Electromagnetic Environment
  - Interference Monitoring
  - Access Point Management

- Environment of Care
  - Temperature and Humidity
  - Pressure Flow (Isolation)
- Real-Time Location
  - Workflow
- Resource Availability and Management

Management of Big Data

The Intelligent Health utilizes a shared seamless medical data environment extracting data from a network of complex medical systems which includes integrated medical devices, clinical applications, and supporting devices. This enhanced patient care environment produces large volumes of data coming for many sources, which are aggregated, collated, sorted, filtered, and processed based on a dynamic relationship defined through the associations between patient and staff-centric scenarios. The data includes a broad spectrum of data types including clinical/physiological parameters, alarms /alerts, patient demographics, and analytics. An Intelligent Health System must provide the capability to process this information in many ways and deliver the information to a broad base of users, including physicians, nursing, administration, biomedical engineering, supply managers, administration, etc. based on the needs of the user.

Cybersecurity

One of the newer challenges and concerns focuses on cybersecurity including data access and attack vulnerability. These complex systems include medical devices, software applications, and IoT and there are two main scenarios of concern which need to be considered. First is a direct attack targeted at a specific device or medical system attempting to disrupt the systems operational functionality. This clearly has a direct impact on patient safety and the has potential of shutting down specific operations, such as physiological monitoring, the ability to perform medication delivery (infusions) or disrupting treatment modalities, and disabling linear acceleration operation and radiation therapy. These types of attacks are commonly focused to disable or stop clinical processes and operations and triggering a ransomware attack. The second type of cybersecurity attack of concern is to use medical device as an entry or access point into the larger health systems infrastructure, which then enables cyberattacks targeting any of the applications, including clinical, business, financial, etc. now interconnected connect through health systems network architecture.

This places new demands on the design of an Intelligent Health System, to ensure these complex systems are accurately evaluated or assessed to define specific vulnerabilities and define specific controls to minimize the potential of these types of attacks. This is a complex and tedious process and typically involves a significant amount of planning and budget to implement. The ability to accurately identify and assess these systems requires a comprehensive inventory process defining the complex systems as well as detailing the supporting integrations to clinical applications, server definitions, etc. Maintaining inventory accuracy to ensure current assessments and control implementation is vital to ensure all incoming devices and system components undergo a cybersecurity assessment and follow traditional incoming regulatory processes,such as acceptance testing prior to deployment. The Intelligent Health System will also need to support an advanced cybersecurity monitoring process or applications, continuously scanning and assessing the infrastructure, servers, firewalls, and devices for potential attacks. These systems will generate alerts and alarms resulting from an unexpected activity, which will also require additional staffing to assess, investigate, and respond to these issues.

Evolving 3D Technologies Enable Patient-Specific Clinical Treatment Solutions

Today Intelligent Health System focuses on advanced treatment modalities which are patient specific and uniquely targeted. These treatments focus on a variety of technologies focused on delivery methodology and enhanced patient safety. These new and evolving technologies include 3D imaging, reconstruction, and printing which have enabled new patient-specific anatomic visualization and opened the door for developing advanced patient-specific treatment solutions. By exploiting this patient-specific anatomic data obtained from medical imaging systems, such as CT, MRI, and ultrasound, an Intelligent Health System can develop customized treatment solutions based on a patient's specific anatomy.

In recent years, new imaging modalities and reconstruction applications have been adopted within the healthcare environment to provide both static and dynamic 3D visualizations of anatomic features. When coupled with technologies and applications such as Medical 3D Printers or and virtual and augmented reality (AR/VR/XR) applications have enabled new capabilities visualizing, planning, patient treatment, and reconstruction. These advances have enabled physicians and clinical staff to develop solutions targeting individualized patient-specific issues and augmenting treatments to enhance clinical outcome. These solutions include

- Patient-Specific Anatomical Accurate Models
  - Advanced Visualization and Surgical Planning
- Custom Surgical Tools and Cutting Guides
- Dental Guides and Tools
- Patient-Specific Immobilization Devices
- Custom Radiation Shields
- Targeting Devices

Surgical 3D Printing Applications

Surgical uses of 3D printing-centric therapies have a history beginning in the mid-1990s with anatomical modeling for bony reconstructive surgery planning. Patient-matched implants were a natural extension of this work, leading to truly personalized implants that fit one unique individual [14]. Virtual planning of surgery and guidance using 3D-printed, personalized instruments have been applied to many areas of surgery including total joint replacement and craniomaxillofacial reconstruction with great success [15].

To develop patient-specific 3D prints, digitization of a patient's anatomical structures must first take place. This leverages standard 3D scanning techniques such as MRI, X-ray CT, or 3D ultrasound to produce a volumetric image of the anatomy. The image components are labeled, via a process called segmentation, to isolate structures of interest and translate the digital information into an STL file which is interpreted by the printer software to create a 3D computer model. The 3D anatomic images can additionally be augmented to drive surgical reconstruction processes. The 3D printer then prints/constructs an accurate full or scale version of the anatomic structures. Medical 3D printing has enabled anatomically accurate physical replicas or models to be fabricated, enabling new visualizations of anatomic structures as illustrated in Figure 2.8. Printer technology enables models to be fabricated using various materials, colors, and even

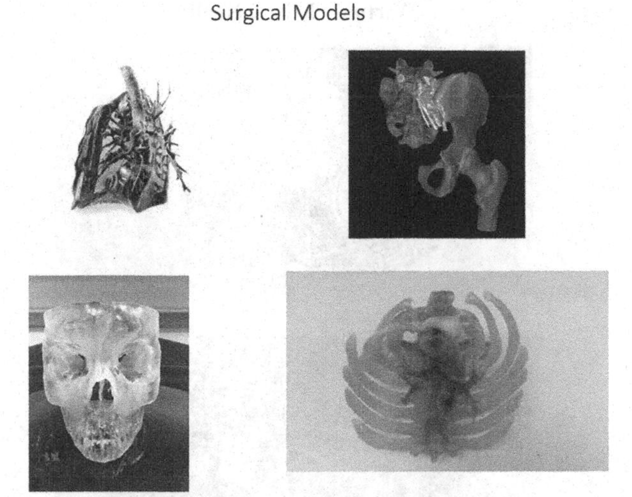

Figure 2.8 Anatomically accurate surgical models.

textures to highlight specific anatomic features, such as veins, arteries, airways, and tumors.

As virtual reality (VR) and augmented reality (AR) systems have become more common place, their applications to the field of medicine have adapted rapidly. Collectively known as "XR," these alternative reality systems, similar to 3D printing applications allow for 3-dimensional patient scans (MRI, CT, etc.), to be visualized in 3D rather than in 2-dimensions on a computer screen. Clinical software products integrate directly with Image Archive Computer Systems (PACS) to pull raw DICOM image stacks and convert them into a 3D hologram for display on a headset such as the wireless Microsoft Hololens. This holographic viewer is transparent, which allows the hologram to be displayed in the space where the clinician is physically located. Using intuitive hand and body movements the user can then access menus, virtual tools, or manipulate the scan by scrolling through slices, rotating the hologram, or simply changing their viewing angle with a head movement. In addition to the 3D representation, the scan can be registered to the patient anatomy, aligning the image to create an overlay with adjustable transparency locked to the patient (Figure 2.9). This allows the clinician an opportunity to view the scan *in situ*, giving anatomical surface references and guidance for surgical planning and has significant potential to impact surgical decision.

Virtual / Mixed Reality

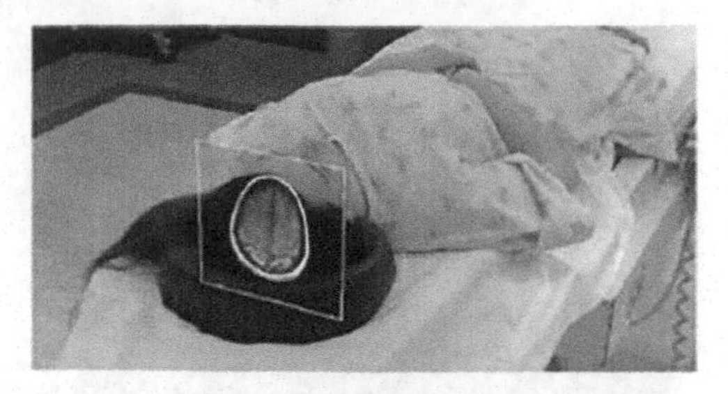

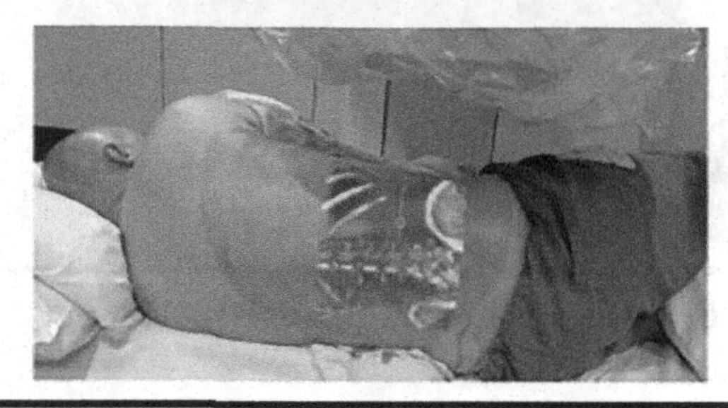

Figure 2.9 Virtual/mixed reality.

Conclusion

As Intelligent Health Systems continue to develop and exploit, integrate, and incorporate new and evolving technologies into the patient care environment and clinical and business processes, new applications will continuously improve patient outcomes and enhance patient care, safety, and satisfaction. The subsequent chapters in this book will discuss and focus on the many components and technologies which have evolved and impacted the Rise of the Intelligent Health System.

Other books in the Intelligent Health Series will focus on the many specific components which have driven and influenced the Intelligent Health System.

References

[1] DeGroot, H., Patient classification system evaluation, Part I: essential system elements. *J Nurs Adm* 1989; 19(6):30–35.

[2] Van Slyck, A., Johnson, K.R., Using patient acuity data to manage patient care outcomes and patient care costs. *Outcomes Manage* 2001; 5(1):36–40.

[3] Arvantes, J., Emergency Room Visits Climb Amid Primary Care Shortages, Study Results Show, AAFP News, 8/27/2008, http://www.aafp.org/online/en/home/publications/news/news=now/health-of-the-public/2008

[4] Frisch, P., Pappas, H.H., *Leveraging Technology as a response to the COVID Pandemic*, Routledge Publishing, 2023, ISBN:9780367769338, DOI: 10.4324/b23264
[5] Morgan, S., Medication error statistics. *The Prescription* (July 2005); 1(1).
[6] Cardinal Health: Statistics – Medication Safety & Education.
[7] Joint Commission Patient Safety Goals, http://www.jointcommission.org/PatientSafety/NationalPatientSafetyGoals/09_hap_npsgs.htm
[8] Dyro, J., *Clinical Engineering Handbook*, Academic Press, 2004, ISBN:012226570X ISBN13:9780122265709
[9] Yadin, D., Bronzino, J.D., Neuman, M.R., von Maltzahn, W.W., *Clinical Engineering*, CRC Press, 2003.
[10] Carlo, B., Cerutti, S., Vienken, J., *Medical Devices*, Springer Cham, Feb 2022, SBN: 978-3-030-85653-3
[11] Frisch, P.H., Miodownik, S., Booth, P., Lui, W., Design of an enterprise physiological data and clinical alarm management solution. Proceedings of the 28th Conference on IEEE Engineering in Medicine and Biology. 2006; 109–112.
[12] Frisch, P.H., Miodownik, S., Booth, P., Carragee, P., Dowling, M., Patient centric identification and association. Proceedings of the 31th International Conference on IEEE Engineering in Medicine and Biology Society. 2009: 1722–1725.
[13] Booth, P., Mehryar, M., Frisch, P.H., Analyzing Clinical Operations via a Patient Experience Visualization Graph. HIMSS 2019 Conference on Intelligent Health Association Handbook 2019, Feb 2019.
[14] "3D Printed Clothing Becoming a Reality". Resins Online. 17 June 2013. Archived from the original on 1 November 2013. Retrieved 30 October 2013.
[15] Fitzgerald, M. (28 May 2013). "With 3-D Printing, the Shoe Really Fits". MIT Sloan Management Review. Retrieved 30 October 2013.

Chapter 3

Visibility in Healthcare with IoHT

Fawzi Behmann

Introduction

The healthcare industry shows great promise as IoT-driven systems and applications are improving access to care, increasing the quality of care, while, at the same time, reducing its overall cost. Today, it is one of the fastest-growing IoT sectors, with a large number of startups developing new medical sensors, transporting the data to care providers, and achieving the desired health improvement outcomes.

It has been estimated that 40% of the global economic impact of the IoT revolution will occur in healthcare, more than any other sector. And IoT-driven companies can gain a competitive edge in that sector – specifically in areas such as user experience, operational costs and efficiencies, and global expansion.

IoT Platform

Today's advanced IoT platforms include a cloud-based software solution for connecting to and managing the connected features of IoT devices.

Service Flexibility – IoT solutions need to address a key requirement of every global business: regional flexibility. A platform must allow

DOI: 10.4324/9781032690315-3

deployment of a global program across a multitude of public and private clouds in a manner that maximizes operational efficiency and maintains local flexibility and autonomy while also ensuring compliance with local regulations.

Customer Experience – When it comes to IoT solutions, customers require and expect ongoing, long-term reliability without the need for constant human touch. Support needs to be agile and responsive to customer concerns and needs. And the platform itself needs to make IoT deployment management as simple as possible, thereby saving work hours and resources.

Modern platforms address these issues with a services delivery architecture that optimizes an end-to-end system – from wireless connection to telematics applications – while simultaneously addressing key concerns around hardware power management, cost structures, and visibility.

Securing an IoT platform requires an end-to-end approach, from physical devices and sensors to data connections, to host systems, to the services, and data stored in the cloud. While security risks can never be completely eliminated, mitigating risks require tools and expertise and development of IoT applications

Solution providers can provide an end-to-end modular technology platform that enables customers to make the pivotal change from unconnected products to connected service offerings.

IoT and Healthcare

The healthcare industry shows great promise as IoT-driven systems and applications are improving access to care, increasing the quality of care, while, at the same time, reducing its overall cost. Today, it is one of the fastest-growing IoT sectors, with a large number of startups developing new medical sensors, transporting the data to care providers, and achieving the desired health improvement outcomes.

It has been estimated that 40% of the global economic impact of the IoT revolution will occur in healthcare, more than any other sector. And IoT-driven companies can gain a competitive edge in that sector – specifically in areas such as user experience, operational costs and efficiencies, and global expansion.

Cellular connectivity and IoT solutions enable medical device manufacturers and healthcare providers to achieve the highest levels of patient engagement and medical adherence, with the lowest total cost of operation (TCO), regardless of global location.

IoHT Devices

Internet of Health Things (IoHT) healthcare devices make precise diagnostics and the monitoring of ongoing vital signs available to those who need it the most: vulnerable patients with seniors with mobility issues, as well as those who suffer from chronic diseases like diabetes, heart diseases, epilepsy, or others.

Wearables measure such vital signs as heart rate, respiratory rate, motion, and electrodermal activity, and send this data to physicians. All this contributes to personalized medicine where patients get only the diagnostics and treatment they need.

Evolution of IoHT in Healthcare

IoHT is promised to improve the quality of service and dramatically reduce healthcare costs. IoHT is already in some parts of healthcare, but it has much more potential to radically change hospitals and medicine.

An interface with IoHT devices will help generate data where insightful information can be gained. Viewing patient information and diagnosing in real time can help improve outcome and patient care experience.

There are several products that have been commercialized including examples such as connected inhalers, glucose monitoring, ingestible sensors, and others. Increasing acceptance of IoHT has the potential of emerging lifesaving applications.

The healthcare industry has just begun to understand the tremendous potential and benefits that can be offered by IoHT through the provision of healthcare devices, services, and interactions.

Medical Adherence

According to research by the World Health Organization (WHO), the benefits of medications used to fight disease are not fully realized because close to 50% of patients do not adhere to medicinal intake guidelines. Reasons for not taking medicines on a regularly needed basis are plentiful, running from lack of funds to sub-optimal healthcare literacy to communication/language barriers to just plain forgetfulness.

IoHT offers a solution using digital pill system to improve medication adherence. A pill with embedded sensors is swallowed by an individual and sends data to the mobile app, helping them track and adjust the medication-taking process.

Use Case: As an example, Wisepill is a leading provider of medical adherence management solutions and the creator of the Wisepill dispenser, a pillbox that uses cellular and IoT technologies to provide real-time medical management solutions. The pillbox, designed to work in diverse environments, has a rechargeable, longer-life battery, which allows the device to be used for extended periods without the need of an external power source.

Patients in developing countries or in hard-to-reach rural areas cannot travel easily to far-off clinics. Additionally, many places have a severe shortage of medical professionals. With IoHT connectivity, Wisepill enables clients, pharmaceutical businesses, doctors, and healthcare organizations around the world to improve medication adherence management. The combination of an experienced IoT solution provider and Wisepill provides patients with the peace of mind from knowing that if they miss taking their medications, there would be a reminder to maintain their medicinal intake schedule. By continuing to apply new cost-saving IoHT technologies, and leveraging economies of scale, Wisepill is providing affordable adherence solutions, assisting millions of people, regardless of where they live.

Blood Banks

Blood units represent a critical aspect of healthcare. Yet, blood units often get wasted due to the inability to store them under appropriate conditions.

The principal goal of an IoHT technology-driven blood bank management program is to optimize the effectiveness of a blood bank. A successful program involves increasing awareness about best practices; reducing the likelihood of blood samples becoming unusable; minimizing blood loss; improving blood availability; continuously educating clinicians; and standardizing operations through workflows.

For example, in India, the business case for an IoHT-enabled blood bank monitoring solution rests on the following goals:

- Monitoring blood bank refrigerators on a 24 × 7 basis and storing relevant data.
- Alerts of temperature variance outside a set range.
- Use of transparent monitoring network (single pane of glass).
- Reduction of paperwork.

The IoHT-based blood bank improvement program includes both a measurement of how well the program meets its goals and also demonstrates a commitment to data-driven reporting. This, plus additional functionality, provides the insights needed to initiate and preserve blood bank management that will save many lives.

Device Monitoring: Defibrillators/Heart Monitors/ Pacemakers

Doctors and hospitals need a secure way to establish connectivity and transmit data from automated external defibrillators to a cloud application. Today, IoHT-connected medical devices can monitor and analyze data coming from the patient in real time. And should an event occur, IoT-enabled defibrillators, for example, provide verbal and on-screen instructions in delivering chest compressions. Some advanced defibrillators even can deliver an electrical shock to a patient's heart. Heart monitors send alerts to both the wearer and the doctor in case of irregularities. In all these cases, device connectivity, with real-time data, literally, is a life and death issue.

With an IoHT-enabled solution for healthcare devices, real-time data, along with alerts and reports, can save lives.

Remote Patient Monitoring

For whole-night heart-rate variability for recovery and readiness analysis, contact-free health monitors have been developed as IoT solutions for healthcare.

1. The remote patient monitoring software connects with ferroelectric sensors to receive a graphical representation of the human body's repetitive movements resulting from a sudden ejection of blood into large vessels with every heartbeat.

The program monitors the heart rate, respiratory rate, and physical activity and records and captures various conditions throughout the night. It then transfers data from sensors to the user-friendly medical application installed on smartphones, tablets, or computers for analysis.

2. Another efficient IoT system (by SeizeIT) aims to measure brain, heart, and muscle activity, as well as respiratory rate and motion. It consists of a patch discreetly attached to the patient's body and a magnet placed on the patch. The solution is used for predicting seizures and enhancing the quality of life of patients with epilepsy.

Remote Care

IoHT has helped transform routine medical check-ups to be more patient and home-centric as opposed to hospital-centric approaches. IoT in healthcare has thus contributed to redefining monitoring, diagnosis, treatments, and therapeutics in customary healthcare viewpoints, thereby reducing costs and errors.

IoHT in healthcare has been embedded in current procedures and systems mostly in the context of remote patient monitoring in real time, collection and transfer of health data from patients outside of traditional medical settings to remotely monitor chronic illnesses.

In the United States, Chronic illnesses account for 80% of all hospital admissions but cost 3.5 times more to treat.

These remote patient monitoring solutions allow greater levels of communication between healthcare providers and patients, ensuring that they receive the information they need before patients experience adverse effects (Figure 3.1).

Telemedicine

An IoHT application is in telemedicine that provides disabled and isolated patients with the opportunity to be remotely consulted and diagnosed by their physicians. In telemedicine, IoT analysis kits are helpful, such as fast self-administered tests. These tests allow patients to self-detect influenza, inflammation, vitamin D level, testosterone, and LH and report these results to their doctors

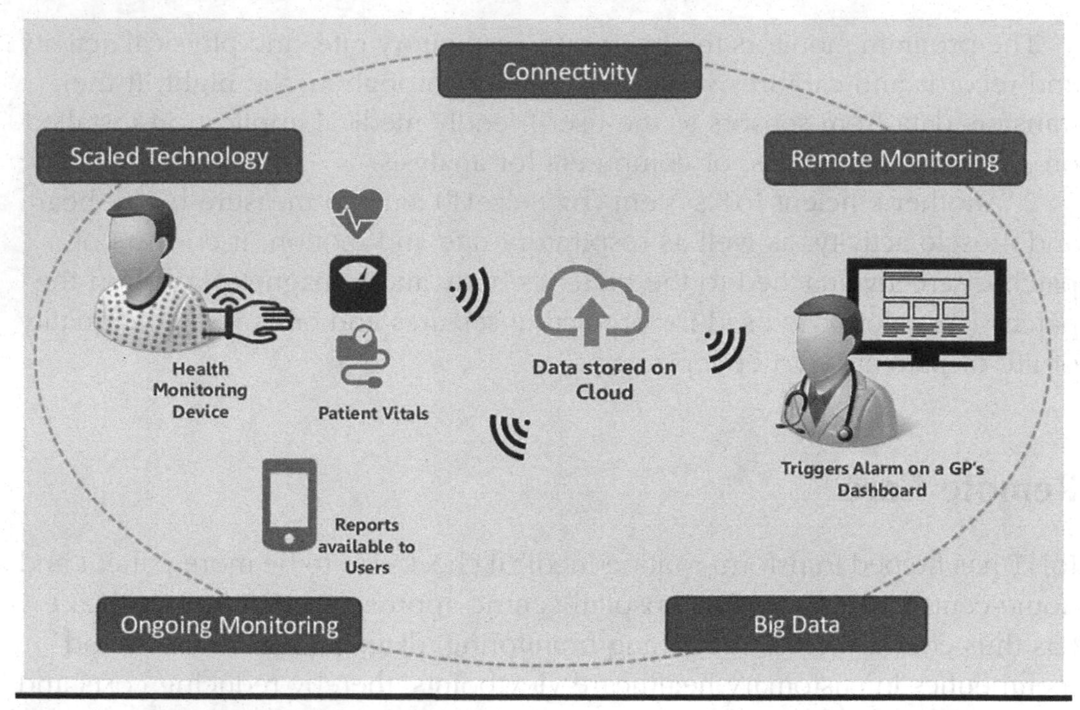

Figure 3.1 Telehealth/telemedicine connectivity.

Caregiver

IoHT in healthcare has begun to pave its path and is employed as a practice among caregivers and patients changing the way patient care was defined in previous decades.

An example provided by SimplyHome offers an IoHT system that communicates with multiple sensors and proactively alert patients and caregivers to changes in behavioral patterns of daily living. The services range from voice-activated environmental controls, personal emergency response systems, GPS watches, motion sensors, stove monitors, and virtual care management. Text, email, or phone alerts can be generated by a single event, an intersection of multiple events, or by inactivity (Figure 3.2).

IoHT and COVID-19

During the ongoing COVID-19 crisis, IoHT has played a pivotal part in properly monitoring patients who are virus-infected through devices and intertwined networks.

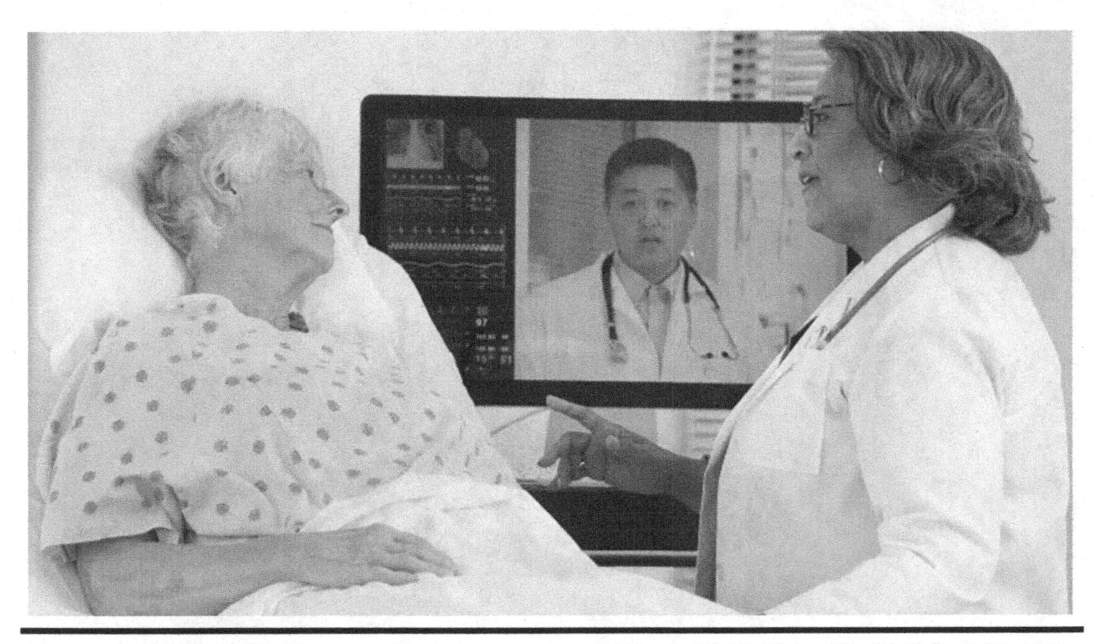

Figure 3.2 Remote telehealth.

The industry has inevitably selected to depend on this communication system to safeguard people against the spreading of the virus. Different applications of IoT in the healthcare sector include Telehealth Consultations, Digital Diagnostics, Remote Monitoring, and Robot Assistance.

Robots

During a quarantine, robots play an important role in keeping medical staff away from isolated patients. For example, robots can be used in different ways, such as capturing respiratory signs and assisting patients with their treatments or food.

Medical Delivery Drone

Medical drones have shown their efficiency at the early stage of COVID-19 where they transfer the COVID-19 test kits, samples, or medical supplies between labs and medical centers to eliminate human interactions. Additionally, this type of drone usually reduces hospital visits and increases access to medical care by delivering medical treatments to patients or another medical center rapidly (Figure 3.3).

Figure 3.3 Medical drone.

Telerobots

Telerobots are usually operated remotely by a human and can provide different services such as remote diagnosis, remote surgeries, and remote treatments for patients with no human interaction during the process. For example, a nurse can measure patients' temperatures without having interactions with them by using these robots. Another example is the DaVinci surgical robot, which is operated by a surgeon while the patient is in the safe isolation of plastic sheeting. This helps to prevent infections by performing surgeries remotely.

Autonomous Robot

Autonomous robots have been widely used during quarantines. They work with fewer or no human interactions and can be utilized in different scenarios in order to sterilize contaminated areas in hospitals, carry patients' treatments, and check their respiratory signs. These will result in decreasing the risk of infection for the healthcare workers while the patients are isolated in their rooms.

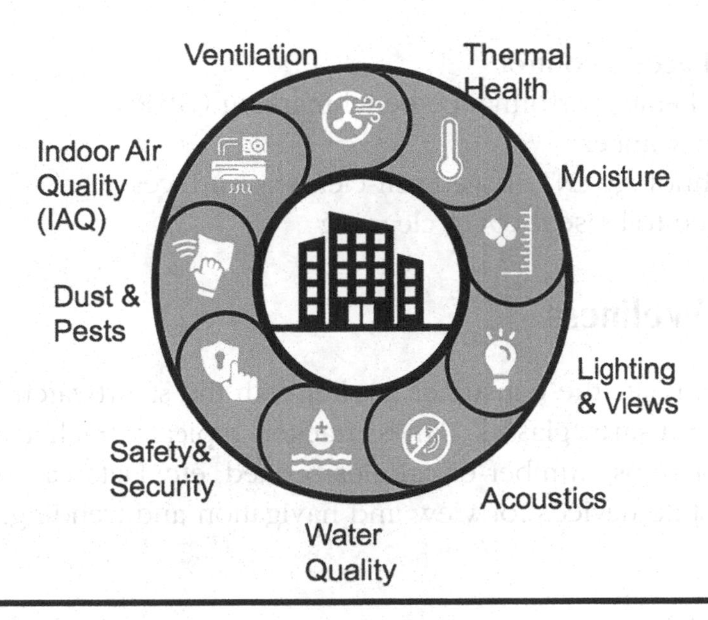

Figure 3.4 Building consideration with COVID-19.

Source: CABA Intelligent Buildings and COVID-19 Report

Buildings are undergoing changes with COVID-19

Due to COVID-19, services have been examined and additional considerations were given to provide a safer and clean environment for work or living (Figure 3.4).

Here is a partial set of services.

HVAC and IAQ

- HVAC optimization
- Indoor air quality (IAQ)-based ventilation
- Ultraviolet germicidal irradiation (UVGI)
- High-efficiency particulate air (HEPA) filters

Post-COVID Applications

- Occupancy detection (OD)-based social distancing
- Indoor positioning (IP)-based contact tracing
- AI-based face and mask detection
- Antimicrobial paints in washrooms
- Foot operated elevator
- Touch-free bathroom fixtures
- Thermal imaging

- AI-based access control
- Building health performance benchmarking (BHPB)
- Pathogen scanner
- Nanotechnology (NT)-based self-cleaning surfaces
- Touch-free toilet seat cover cleaning

IoHT and Wellness

Wearables dominate the consumer market with the smartwatch, activity/fitness tracker, and smart glasses. Fitness trackers typically track activity such as the number of steps, number of calories burned, etc. Data can be communicated to mobile devices for view and navigation and trending.

IoHT and AI

Some of the IoHT advancement include advancements in sensor technologies, improvements in systems to gather and process data, and integration of artificial intelligence (AI) technologies in healthcare.

As described earlier, sensor technologies are improved by the developments in the ability to collect and gather data that eliminates manual data entries, and automation reduces the risk of errors. Smart sensors are a combination of sensors and micro-controllers that will assist IoHT in terms of accurate measurement, analysis, monitoring, and assessing a multitude of parameters in healthcare.

The rise of AI and its alliance with IoT is one of the critical aspects of the digital transformation in modern healthcare. This is likely to result in speeding up the complicated procedures and data functionalities that are otherwise tedious and time-consuming. AI along with sensor technologies from IoHT can lead to better decision making. Advances in connectivity through AI are expected to promote an understanding of therapy and enable preventive care that promises a better future.

Predicting Diseases – Potential Use Cases

Because AI has the potential to store all of people's data in one place, it can access this information to compare patients' previous ailments with the current symptoms to come up with a more accurate diagnosis. Since such apps have millions of the previous diagnosis stored, patients utilizing them don't feel the need to go for a second opinion. Plus, by combining and analyzing data from various sources, AI can predict health problems that a person can get in the future.

One such application is Verily by Google, which started as Baseline Study and is working on forecasting non-communicable diseases like cancer and heart attacks as well as hereditary genetic illnesses. The aim is to enable doctors to predict any condition a person is likely to suffer in the future so that they can come up with treatment plans to avoid it or treat it in time.

Another way AI can help with diagnosis is by identifying biomarkers. Biomarkers are special molecules present in bodily fluids that can locate the presence of a particular disease in a person's body. Because AI can automate the majority of the manual work in recognizing these biomarkers, it can save a lot of time and energy when it comes to diagnosing a disease. Since AI algorithms can skillfully classify molecules to identify a specific condition, they're more cost-efficient as well; people will no longer have to go through expensive lab tests and other procedures like whole genome sequencing if doctors utilize the aid of AI in diagnosing diseases.

AI is now being used in operating rooms because applications based on AI have proved to be a great help to doctors during surgical procedures. Robot-assisted surgery is now a thing of the future in the healthcare industry and has paved the way for successful surgical treatment of rare conditions. Complex operations can now be performed with precision with minimal side effects, less pain and blood loss, and a quicker recovery through robot-assisted surgery, as reported by Mayoclinic.

Along with this, AI has now provided surgeons with real-time information about patients currently under treatment. This includes visual scans revealing the division of the brain into its various portions as well as MRI scans and other imagery required to assist in the operation. By laying these images on top of the patient's body, AI has given X-ray vision to the doctors. This comes as a great relief for the patients who feel more secure when handing themselves over to their physicians under anesthesia.

IoHT and 5G

With the development of the 5G network infrastructure, IoHT technology offers humankind an experience that was never available before. With its ultra-low latency and vast bandwidth, the 5G connection allows lightning-fast

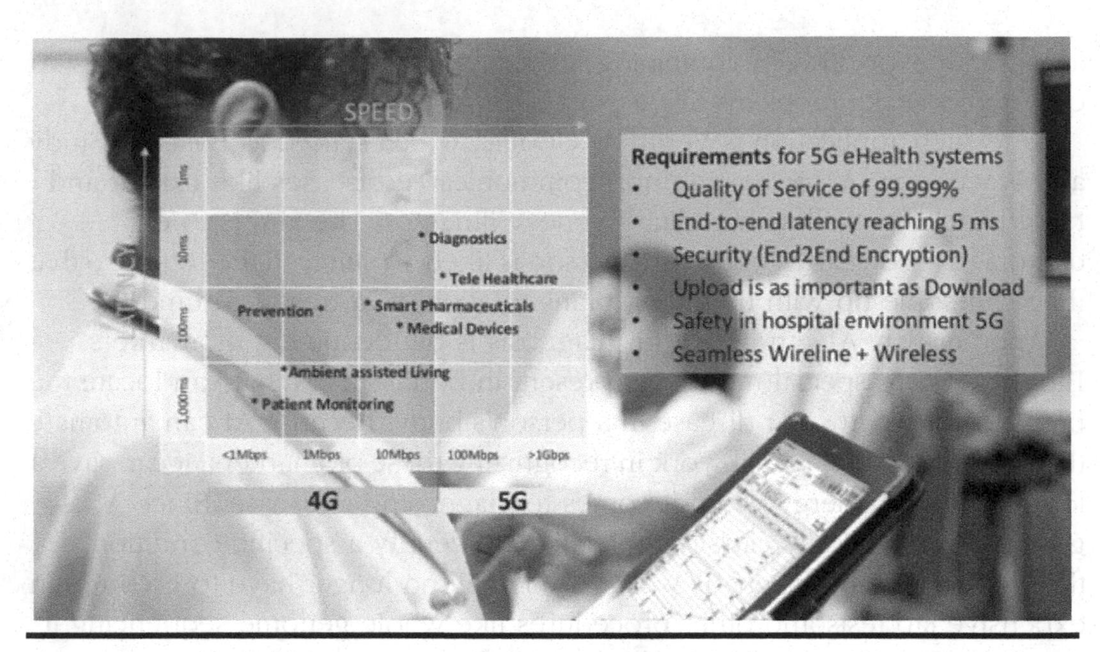

Figure 3.5 5G enable URLLC Remote arm surgery.

heavy data exchange like MRI images. It opens up the potential for ground-breaking treatment such as remote robotic surgery.

Thanks to sensors placed on the robot and the virtual reality headset, this enables the surgeon to perform real-time surgeries while being miles away. This technology promises access to high-quality treatment to a greater amount of people from all over the globe (Figure 3.5).

IoHT and Cyber Security

Prevalent challenges revolve around the generation of tremendous data through a high number of devices associated with the system and threat of cyber attacks and data breaches.

Despite the exciting advancements offered by IoT and AI in healthcare, the multitude of information collected in the forms of digital pathology data, diagnostic data, sensor data, EHR data, imaging data, and others leads to excessive deposition of data that are difficult to handle. This could lead to unauthorized access of this data by cybercriminals to create fake IDs, smuggle drugs or file a fraudulent insurance claim on the patient's name. Solutions for data scalability and protection through cybersecurity would address the critical nature of IoT in healthcare boosting its applications and adoption rates in the future.

Conclusion

IoHT in healthcare is already at the door. It's becoming a medical world reality and affects our lives already, altering the understanding of what cutting-edge diagnostics, treatment, and healthcare look like today.

There are limitless predictions about the revolution that can be brought through the IoHT in healthcare by improving the quality of healthcare and dramatically reducing healthcare costs. Taking a closer look at the technical aspects, the role of IoHT in healthcare is yet to be explored at greater depth with regard to smart sensors and data analytics.

IoHT in healthcare is expected to bloom and overcome its challenges to revolutionize the conventional healthcare models of the future.

Chapter 4

Medical Device Security Program for a Healthcare Delivery Organization

Ali Youssef

As healthcare institutions around the world evaluate leveraging technology to improve patient care, the vast majority of devices and platforms under consideration leverage microprocessors and rely on being connected to some form of network. With the latest advances in AI and ML, there is a dependency on cloud based as well as on-premise Infrastructure. Medical and IoT devices play a prominent role in the automation and efficiency of patient care in the hospital as well as in the home care setting. These devices are increasingly being used in mission-critical and in some cases life-critical settings even though it is widely known that they often lack a fundamental cybersecurity focus. Healthcare delivery organizations (HDO) look to medical device manufacturers to design and adequately test the clinical efficacy and safety of medical devices while healthcare technology management departments are busy keeping up with mandates from the joint commission mainly around preventative maintenance and emergency management.

Over the last several years, there has been a realization from healthcare technology management teams that cybersecurity events can disrupt medical devices and directly impact patient care and safety so departments around the country are beginning to pay attention to this. Hospital administrators are concerned from a patient safety standpoint as well as the possibility of

 DOI: 10.4324/9781032690315-4

tarnishing the reputation of their institution. In many cases, the teams are not adequately trained in IT and cybersecurity which leads to a joint effort in tackling this space between IT and HTM departments. To make things even more convoluted, Information Privacy and Security teams do not generally report to the IT CIO to avoid a conflict-of-interest scenario where IT departments are auditing their work.

This chapter intends to provide a framework for an HDO to establish a medical device and IoT security program. The medical device design and market release process is heavily regulated by the FDA, but unfortunately, the process of establishing a medical device and IoT security program within an HDO lacks any centralized regulation or enforcement. The silver lining is that many organizations like AAMI, NIST, and others have developed guidance and elements of a construct and a roadmap to build these types of programs. Some large health systems have mature medical device security programs but the vast majority are at the beginning of their journey.

Medical devices are becoming more advanced, relying on the Internet, hospital IT networks, and mobile devices to function properly and share information. It is becoming increasingly important to focus on the security of these types of devices to maintain device efficacy and ultimately patient safety. These highly regulated devices are easy targets because they have a long lifecycle (10–20 years in some cases) and are usually not up to date with the latest security best practices in part due to the nature of the FDA 510K process. Interconnectivity and interoperability of medical devices enable important advances in patient care, but these attributes also present security risks. Device security vulnerabilities can adversely affect device performance as well as the availability, confidentiality, and integrity of the device and its data. The Maximum Tolerable Downtime for the health system decreases from days to hours when medical devices are compromised and HDOs often do not have mature business continuity plans focused on medical devices. The responsibility for medical device security is a shared one between the device manufacturer and HDO. There are four interdependent lifecycles including medical device manufacturer development lifecycle, medical device manufacturer supply chain, medical device manufacturer maintenance, and Health Delivery Organization. The program outlined in this chapter is focused on the responsibilities of the HDO. A breakout of the medical device lifecycle in an HDO is illustrated in Figure 4.1.

The medical device security program is aligned with traditional Confidentiality, Integrity, and Availability cybersecurity objectives but requires additional patient care and safety considerations. Proposed security

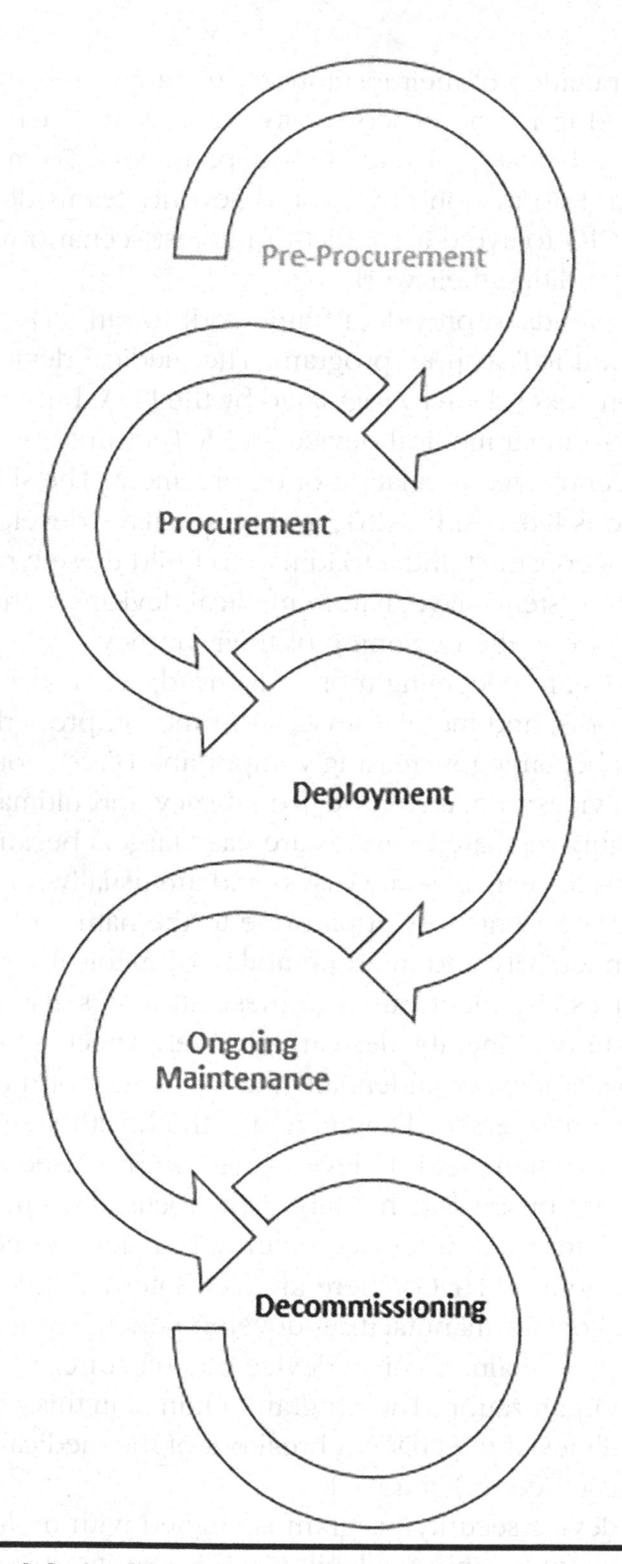

Figure 4.1 Medical device lifecyle in a healthcare delivery organization.

measures are sensitive to not interfere with the device's intended usability or patient safety. Ultimately the program is intended to reduce negative impacts on the delivery of care and to avoid the loss of sensitive health information.

Several security frameworks can be and have been used by IT departments to address cybersecurity. These include the NIST CSF, ISO 27002 NIST 800-53, or SCF. One of the most popular and widely adopted in HDOs is the NIST Cybersecurity Framework which is a subset of NIST 800-53 and relies on certain elements in ISO 27002 but does not include all the elements in both.

The Program

The medical device and IoT framework being presented in this chapter along with its 12 focus areas are primarily aligned with the NIST CSF. This is one way to tackle this space but depending on your institution there may be a need to modify or augment it. The focus areas are aligned with best practices and attempt to leverage existing groups and teams as much as possible to reduce the need for a large medical device/IoT security team. In most cases, the existing security policies, processes, and procedures need minor modifications and rewording to address medical devices. The 12 focus areas for the program are outlined in Figure 4.2.

The focus areas are interconnected and, in some cases, have dependencies. As we work through describing each one, these will be addressed. One chapter is not sufficient to do a deep dive into each of these, but the core premise and considerations will be discussed.

Asset Inventory and Management

The process of quantifying the risks introduced by IoT and medical devices to the business relies on two foundational elements. These are attaining a thorough understanding of the inventory of connected medical devices and when new devices are being introduced into the environment. To manage the risks associated with medical devices, we must first have an accurate inventory of medical devices. There are usually several stakeholders within the health system that maintain their medical device inventories. These can include HTM, Radiology, Lab, the Simulation Lab, and others. The largest documented medical inventory is usually managed by HTM and housed in the Computerized Maintenance Management Systems (CMMS). The CMMS

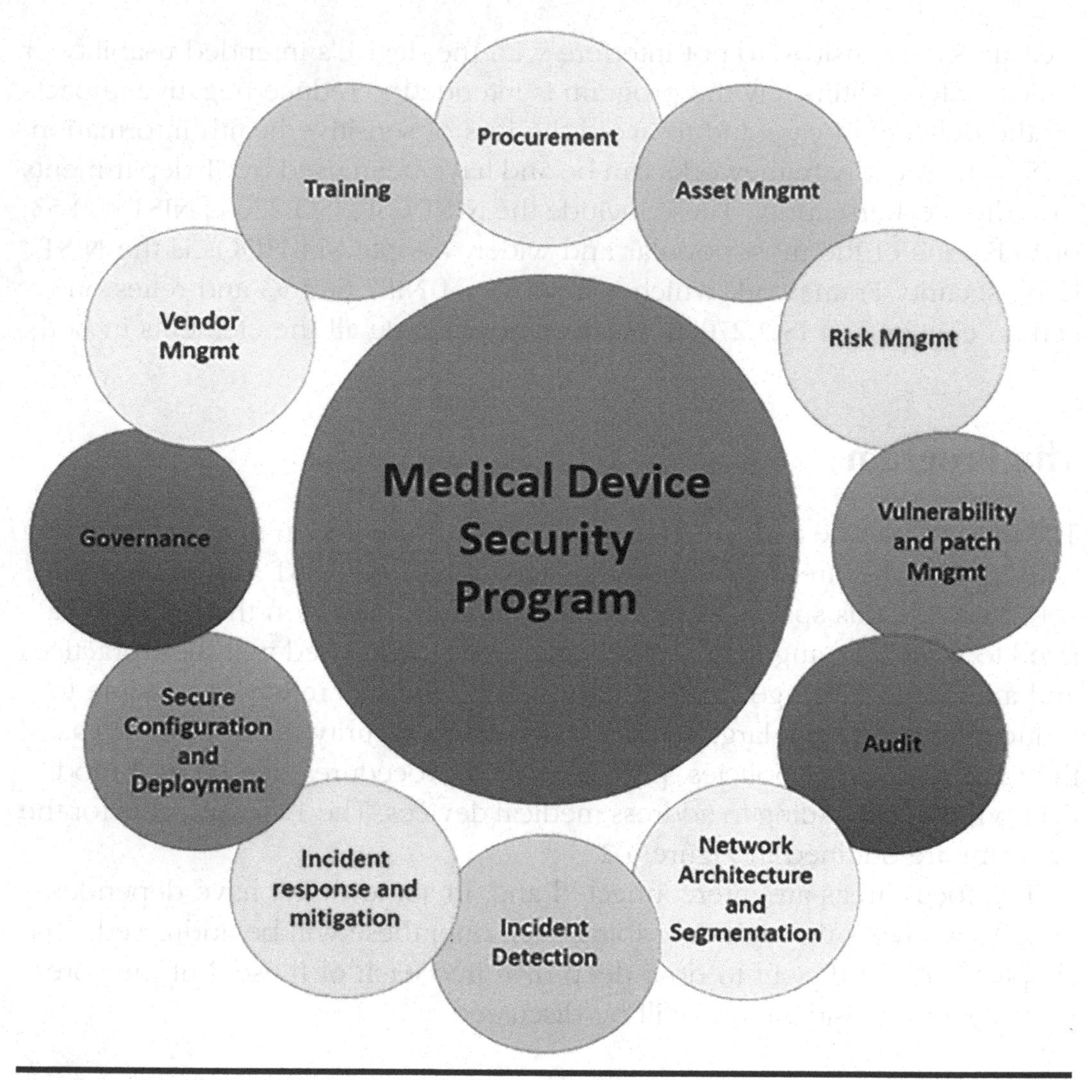

Figure 4.2 Focus Areas of Medical Device and IoT Framework.

data does not typically include IT, or security-related elements and is not tied to the CMDB that is used by IT.

IT and CE see the world through their own biased lens. IT is focused on mission-critical assets while CE is primarily dealing with mission/life-critical assets. In a typical healthcare facility, approximately half of medical devices are networked, so the two groups have a growing inter-dependency. The configuration management database (CMDB) is aligned with ITIL principles, but the CMMS is not. It is designed for medical equipment lifecycle management and the data elements it is focused on reflect that.

To improve the inventory alignment between the IT network scanning, network access control tools, and the CMMS inventory, the data elements outlined below must be captured in the CMMS inventory record. HTM can integrate collecting this information into their existing medical equipment onboarding policies. The additional information collected will help reconcile the inventory and provide insight into vulnerability and patch management as well as risk management.

The key information associated with each medical device should be documented in the medical device inventory. This includes the following:

- Unique identification number
- Type of device
- Device description
- Device manufacturer
- Device model
- Device serial number
- Device department ownership
- Device physical location within healthcare facility
- Device acquisition date
- Installation date and acceptance testing information/results
- Device cost (original and replacement)
- Device condition/operating status and rationale (e.g., in-service or out-of-service)
- Operating and power requirements
- Source of maintenance services (e.g., contract or time-and-materials vendor or in-house)
- Scheduled maintenance procedures (e.g., items to check, calibration to perform, parts to replace) and frequency
- Record of maintenance performed (e.g., dates, service provider, activities performed, test results/findings)
- Corrective maintenance (e.g., description of problem, source, and cause of problem such as scheduled maintenance report or manufacturer recall, description of actions taken, service provider, parts/time used, post-maintenance status of device)
- Records to facilitate replacement planning (e.g., current physical condition, utilization level, status of technology, operating reliability, availability of manufacturer and/or third-party support)
- A score associated with criticality of the device (e.g., whether failure of the device could result in loss of life, serious/permanent injury, minor/temporary injury, or negligible/no injury)
- A score associated with probability of maintenance-related failure

- A maintenance-related risk score, which is a function of the criticality and the probability of maintenance-related failure
- A description of any stored or transmitted data types (e.g., PII, PHI, clinical results)
- The type and version number (both that is currently on the device and the latest known available) of the operating system, software, and firmware
- Whether the connection is wired, wireless, USB, or proprietary, if applicable.
  - List of interfaces: USB, Bluetooth, Ethernet, Wi-Fi, HDMI, proprietary
  - MAC address1 (Wi-Fi)
  - Wi-Fi chipset
  - Wi-Fi driver
  - Wi-Fi firmware version
  - Wi-Fi authentication (PSK, 802.1x)
  - Wi-Fi encryption (WEP, TKIP, AES)
  - Wi-Fi Physical layers (802.11a/b/g/n/ac/ad)
  - MAC address (Ethernet)
  - Additional Ethernet or MAC addresses
  - IP address (if fixed)
  - Bluetooth address
  - Bluetooth chipset mfg. and model
- The hostname (the logical name assigned to network devices, typically servers)
- The IP (static or dynamic) and MAC addresses if networked (or networkable)
- The built-in security features from MDS2 – Manufacturer's Disclosure Statement for Medical Device Security
- The device's configuration information and Operating System (current preferred settings on features)
  - Operating system (iOS, android, windows, Linux, etc.)
  - Operating system version
  - Other potentially relevant software versioning information:
  - Operating system patch level
  - Bootcore software version
  - Application software version
  - Firmware version
  - Antivirus software vendor and version

    - Antivirus database version/date
    - Other installed malware protection (e.g., HIDS/HIPS)
    - Versions of various libraries in use on the device.
- The dependencies (parent/child/sibling connections)
- The peripherals connected (e.g., screens, printers, storage devices, associated devices, and systems)
- Authorized remote access (description of connections and authorized parties for authorized remote access)
- Record of cybersecurity maintenance performed
- Cybersecurity bill of materials (e.g., commercial open-source, and off-the-shelf software and hardware components)
- A score associated with severity of the device (e.g., whether cybersecurity compromise could result in loss of life or significant loss of protected health information or other sensitive data; permanent adverse effect; reversible adverse effect; no or negligible adverse effect)
- A score associated with probability of cybersecurity-related failure
- A cybersecurity risk score, which is a function of the criticality and the probability of cybersecurity-related failure
- Date inventory performed/updated

Reconciling a medical device inventory and managing security manually is not scalable or sustainable. The health system would be better served by introducing an automated mechanism for detecting and managing medical devices on the network. Several platforms can help bridge the gap between the CMDB and the CMMS as well as automatically populate the new data elements within the CMMS. This type of platform can enrich the data in both systems and be used as a centralized vulnerability and risk management tool. It can also tie into the network access control platforms to enforce device micro-segmentation as needed.

Procurement

The focus on procurement is intended to proactively capture security-related information about medical devices as they make their way through the sourcing and procurement process. The information provided by the vendor will be used as one input into assessing the risk of introducing the device onto the network. The goal of the medical device procurement process is to ensure that HTM, IT, and Security are allowed to review a medical device before allowing it onto the network. This is intended to manage the risks associated with new medical devices being added to the network and to avoid adverse interactions with other networked devices.

Many departments and teams can purchase medical devices, but the requests must make their way to the central sourcing department. To assess the risk associated with introducing a medical device onto the network, a series of documents is required from the medical device vendors. These include

- A MDS2 document that details the device's cybersecurity features and vulnerabilities.
- A manufacturer residual risk file.
- The software/firmware BoM.
- Any recommendation from the manufacturer regarding optimum security configuration.
- Detailed technical specifications for the device.

HTM, IT Security, and members of the clinical community can review the responses and craft a risk assessment document back to the business.

Risk Management

There is an inherent risk in everything that we do and ultimately the equation boils down to the probability and severity of a failure or a compromise. We are constantly gauging the probability of an event occurring and the impact it would have if it did. This is no different when it comes to medical devices. The goal is to document the probability of failure and the severity associated with it. The severity of failure associated with a medical device is constant, but appropriate controls can be used to decrease the probability. Four levels of severity can be defined as negligible, marginal, critical, and catastrophic and probability can be broken down into Improbable, remote, occasional, or probable. If one were to assign numbers to each of these beginning with the lowest number for the least probable and severe, the equation of probability x severity will yield a risk score.

When it comes to risk management, it is important to capture the risk associated with existing devices as well as new devices. The documents collected as a part of the procurement process are foundational to performing and documenting a risk assessment. When it comes to existing medical devices, passive vulnerability scanners can shed light on known vulnerabilities and are a great starting point for a risk assessment. Once the risks are identified, a risk register needs to be created to track how they are being addressed. IT and HTM can shed light on certain risks, but at the end of the day, clinicians will need to determine how these should be addressed. The business should be presented with a risk mitigation plan along with any associated costs. The risk mitigation involves using compensating administrative, technical, and physical controls. Figure 4.3 outlines some examples of each.

The findings from the risk assessment process should be documented and presented to the medical device security committee and tracked to completion. If the business determines that the risk is acceptable, this should be documented to ensure that it is clear that the risks were considered and collectively reviewed.

Following any mitigation, the residual risk should be documented and re-evaluated at regular time intervals. Risk is constantly changing based on emerging vulnerabilities and configuration management.

Administrative Controls

- Policies, procedures, training
- Business Associate Agreement
- SLA and SLO with manufacturer

Physical Controls

- Locks
- Alarms
- Secure Location

Technical Controls

- Network based security measures
- Device based security Measures
- IPS/IDS
- MFA
- Antivirus

Figure 4.3 Controls.

Vulnerability and Patch Management

Mature IT security programs generally have a high level of automation and a good handle on vulnerability and patch management. This is due to years of practice leveraging vulnerability scanners and dealing with sophisticated vendors that have streamlined the release of new firmware patches. Unfortunately, medical and IoT devices do not share this luxury. Manufacturers are incentivized and driven by the clinical efficacy of the devices, but they generally are not concerned with the underlying operating system, threat modeling, or security concerns. The design engineers will use the most convenient operating system which could be one that is legacy or no longer supported. A good example is Windows XP, which has been unsupported by Microsoft since 2014. To make matters more complex, these types of devices generally do not react well to traditional vulnerability scanners like Qualys, Rapid 7, etc. The act of scanning a given device can impact its core functionality. The trends are concerning enough that medical device test labs at healthcare institutions will not pull inventory out of the clinical environment to test. There is an apprehension that scanning a given device in the lab can impact its clinical functionality. In some cases, medical device manufacturers will void the warranty on a device if it is scanned by a vulnerability scanner without their knowledge or approval. If your institution intends to take the risk of going down this path, it is paramount that there is language in the contract with the manufacturer spelling this and the support expectation out. The last thing anyone wants is a non-functional MRI that the manufacturer refuses to support because it was scanned for vulnerabilities and is no longer functioning to expectations.

The most effective and automated method to deal with medical device vulnerabilities is by using a purpose-built medical device passive scanning platform. These intercept and analyze data from the network without interacting with the medical device directly. This type of platform brings vulnerabilities to light and allows traditional vulnerability scanners to avoid medical devices on the network. A huge benefit to going down this path is that the vulnerabilities are prioritized and only the ones that matter to your organization are in focus. Attempting this prioritization manually while keeping up with the emerging vulnerabilities is an unrealistic, herculean task.

A good way to augment this functionality is to establish a medical device lab where more intrusive vulnerability scans can be performed. Vulnerabilities should be tracked in a similar manner to risks, and there needs to be a sufficient time window for remediation. In many cases, it may take medical device manufacturers months to create a patch for a given vulnerability.

Routine patches associated with ongoing preventative maintenance can be addressed by the HTM department based on their needs and the recommendations from the manufacturer. One of the keys to a successful program is for HTM to view security risks to medical equipment through the lens of preventative maintenance. A medical device whose security has been compromised can cause harm and possibly death. To that end, the policies for onboarding, operating, and decommissioning medical devices should include cybersecurity considerations

Audit

Currently, in mid-2022, there are no regulatory organizations that are responsible for auditing medical device security programs in an HDO. The focus is generally on compliance and is privacy centric. The most effective way to gauge the maturity of the medical device security program is by relying on specialized consulting firms and using internal audits. The team can audit that the medical device security policies are being followed and that any deviations or exceptions are documented.

Network Architecture and Segmentation

Network infrastructure has changed significantly over the last 10–15 years. What used to be a simple switch or wireless access point deployment has become much more over the years and now includes network access control, analytics, and centralized management amongst other things. Figure 4.4 depicts this transition.

When it comes to medical devices, consistent and homogeneous network architecture can make the support burden much simpler. If a consistent nomenclature and approach to Virtual Local Area Networks (VLANs) is used, it can make micro-segmentation easier to deploy. The reality is the majority of health systems do not use a consistent design strategy when it comes to medical devices. The architect assigned to each project has a tremendous influence over the design. Network architecture is important for a number of reasons including ensuring there is sufficient bandwidth to support each use

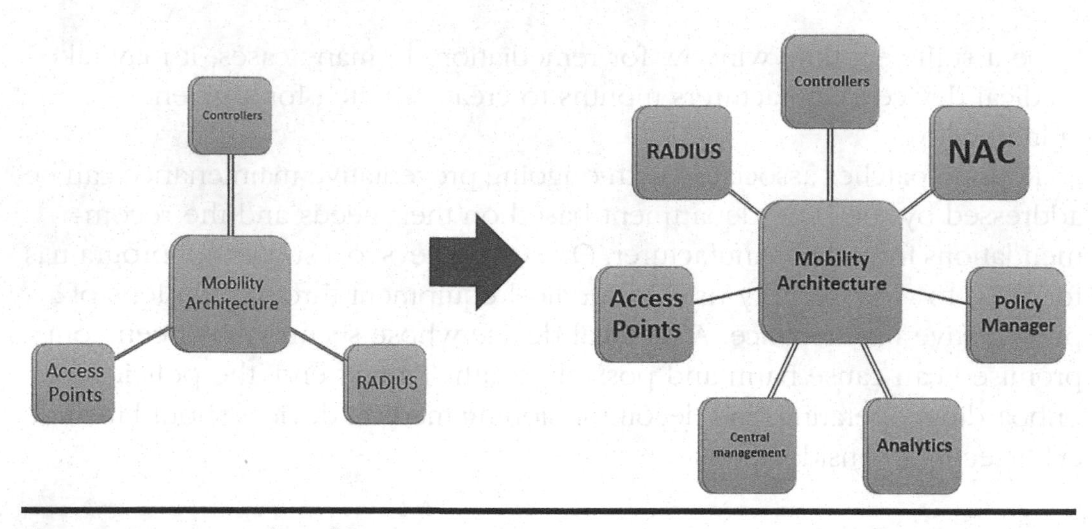

Figure 4.4 Network architecture transition.

case, the ability to segment traffic, and the ability to identify connected medical devices.

In a mature medical device program, network access control plays a central role in protecting medical devices from the network and vice versa. When combined with a medical device security management tool, NAC is an essential component for restricting network access to medical devices down to only what is required. Often medical devices have access to IPs and protocols that are not required.

Another aspect to consider is the need for medical device manufacturers to access their devices remotely for routine maintenance and updates. It is paramount to ensure that the type of remote connectivity aligns with the organizational policy and that can include ensuring that there is an audit log of who is connecting, when they connect, and the detailed changes that are made. This provides an audit log to ensure that every change on the network is captured and logged via the standard change management process which facilitates a much faster response from support staff if issues arise as a result of changes.

Incident Detection, Response, and Mitigation

Incident detection involves identifying when a medical device cybersecurity event or anomaly has occurred in a timely manner. The sooner an incident is detected, the faster the team can investigate and address it. In order for these types of events to be detected quickly, there needs to be an automated system continuously monitoring anticipated medical device behavior and traffic patterns and immediately reporting any irregularities to the security group and other stakeholders. The detection system should be routinely updated with information about new devices entering the environment. Every device will have unique traffic requirements and patterns; any deviation from these should set off an incident. Each incident merits further scrutiny from the security team to determine if it is a viable threat. Audits and tests should be performed at regular intervals to ensure that the detection software is functioning to meet the needs of the hospital and is up to date.

Input into the detection process should include the ability for external and non-technical stakeholders to submit reports, e.g., clinicians, patients, law enforcement, government agencies, security researchers, and vendors. When it comes to response, a suspicious event or anomaly in medical device traffic patterns needs to trigger a standard incident response plan. The response includes a communication plan as well as measures to mitigate the incident as needed. The first step is to investigate the event further to determine its validity. Once a cybersecurity incident has been validated, a standard communication plan involving stakeholders should be followed. The communication strategy will depend on the extent of the incident and any breach identified. Interdisciplinary stakeholders including clinical staff should be notified and engaged in order to prepare for a potential disruption in care and prepare backup methodologies as needed but also to support the response plan with clinical decision making (e.g., when to divert patients). In the case where an incident with a medical device is identified, the medical device manufacturer, as well as the appropriate regulatory agency or authority having jurisdiction, should be notified. In cases where a data breach is identified, the Secretary of HHS (or local Department of Health) should be notified based on current regulations. Public Relations may also need to be notified in order to announce the findings via local media outlets. Other stakeholders would include law enforcement, information-sharing analysis centers (ISAC or ISAOs), and government cybersecurity agencies. An in-depth analysis including forensics and reviewing all relevant and necessary event logs is advised, to understand the full extent of the incident. Mitigation activities should be performed to prevent potential expansion and resolve the incident.

Following the response to an incident, it is crucial to kick off recovery and restore any capabilities and/or functionality lost due to the incident. A standard and tested recovery process needs to be followed to ensure that systems are fully restored as quickly as possible. The last phase of the response aligns well with the ITIL framework. Lessons learned from the incident detection and response should be documented, reviewed, and follow a continuous improvement cycle. This includes reviewing technical improvements as well as policies, procedures, and processes, including the incident response plan.

The plan should be continuously refined, and the team should practice the plan by implementing tabletop as well as functional exercises. Clear communication is a thread that ties detection, response, and recovery together. Key stakeholders, including Risk Management, Legal, and Communications should be kept up to date throughout the process and provide their support as needed.

Secure Configuration and Deployment

Every medical device has unique requirements when it comes to secure configuration and deployment. Part of this is configuring the device appropriately, but the environment and end-to-end communication of the device on the network are of equal importance. A reference architecture and a device configuration guide will shed some light on best practices, but ultimately it is up to the healthcare organization to ensure that the configuration and deployment runbooks abide by their policies and standards.

There are some common themes when it comes to the gold standard for configuration including supporting the right level of encryption and authentication, but what makes this focus area labor-intensive is the fact that each device will need a runbook documenting its detailed configuration.

In addition to the configuration of the device itself, the network infrastructure will need to be modified to cater to the connectivity needs of each device. Let's take an IV pump as an example. If the pump relies on Wi-Fi connectivity, then configuring it to abide by the enterprise wireless standards is important, but it is also important to ensure that the device does not have unrestricted access to the network. The infrastructure should recognize the device as an IV pump and enforce a policy to ensure that the pump can only reach what it needs on the network and nothing more. In order to be successful in this area, the medical device manufacturer will need to provide the required documentation and the requirements of the device on the network. A medical device security management tool can confirm the validity of the information provided by the vendor and in some cases, it can interact directly with NAC to ensure that the appropriate traffic policies are implemented.

Governance

Governance is at the heart of the medical security program. It is an essential part of engaging with all the stakeholders which can range from pharmacy to radiology, informatics, IT, lab, sourcing, and others. Having a clinical executive sponsor sets the appropriate tone around governance. Medical device security is not an IT issue and ultimately it boils down to patient safety and the institutional appetite and threshold for acceptable risks. Governance encompasses two key areas, namely, establishing the appropriate policies/standards, and committees/workgroups for the program. The security team is responsible for shining a light on vulnerabilities and risks associated with medical devices and recommending remediations, but the clinical stakeholders need to choose which direction to take when it comes to addressing risks. Policies that address the medical device lifecycle within the health system have been around for many years, but they often lack specific security-focused language. These typically address pre-procurement, procurement, deployment, operations, and decommissioning of medical devices. An overarching medical device security policy that points to other existing policies is a great way to pull all of the information into one document. The existing policies can be modified to include a security focus. For example, part of the pre-procurement policy can be requesting the appropriate security documentation from the medical device manufacturer so that a risk assessment can take place prior to purchasing new devices. When it comes to operations, the policy will need to include how to deal with emerging vulnerabilities, and ultimately when the devices are decommissioned, any configuration or PHI that may reside on a medical device needs to be scrubbed appropriately. In some cases, this means installing a new hard drive prior to decommissioning a medical device.

From a governance committee standpoint, three tiers of needed to effectively manage the program. These are outlined in the diagram below.

The highest Tier is comprised of senior executives in the organization who report to the board and need to be kept up to date with risks to the institution posed by medical devices. Reporting into the highest tier is a security committee comprised of the various stakeholders including IT, HTM, Lab, pharmacy, clinical specialties, radiology, and others. This committee can meet up to once per month to discuss emerging trends and specific risks being used by different types of medical devices along with recommended remediations. The committee will need to make determinations on how to deal with the risk or in some cases accept it.

Reporting to the medical device security committee is a medical device security workgroup that is comprised of SMEs that meet on a regular basis to discuss the outputs from the medical device security management tool. This is where different types of security controls are recommended and tracked. Only escalations or risks that have complex dependencies or some level of funding bubble up to the committee. Any exceptions or deviations from the policy are addressed and documented at the committee level. The diverse stakeholders are instrumental in ensuring that no policies are missed and that risks are classified and prioritized appropriately.

Vendor Management

The most effective way to manage security expectations from medical device manufacturers and vendors is to include these in the contract language. The Healthcare and Public Health Sector Coordinating Council (HPHSCC) released model contract language for Medtech cybersecurity which outlines a number of clauses including the availability of an SBOM, remote access expectations, etc.

The expectations around SLA/SLOs and reaction times need to be clearly defined. One general pitfall to avoid is trying to predict the level of security maturity of a vendor based on their historical performance. There is not a direct correlation between the two.

The organization should maintain contact information for the security group within each MDM to ensure that they can maintain an open line of communication and that they have the ability to reach out as needed.

Training and Measures for Success

The ultimate measure of success is ingraining medical device security awareness into the organizational culture. That means staff members should be on the lookout for concerning anomalies with medical devices and these should be reported to the HTM and the security group for further investigation.

Routine, cross-functional training is required in order to attain this level of maturity. That means training IT on HTM concepts and vice versa as well as training the business on the importance and the risks associated with medical device security. Medical device safety can impact patient safety directly which is one of the most important measures for an effective health organization.

ROI Indicators

One of the first questions an HDO administrator will inquire about when presented with the recommendation to establish a medical device program is about the ROI of such a program. A perception exists that MDMs should be responsible for this space, with little involvement or investment from the HDO. This perception is flawed because the security of a medical device depends in part on the environment it is used and its configuration. Imagine an intravenous pump that has the capability to be configured to use the gold standard of authentication and encryption. This would be meaningless if the end user (e.g., HDO HTM, and information technology staff) does not configure it appropriately or cannot support that level of security on their network.

ROI can be classified as "soft ROI" (e.g., new patient attraction/existing patient retention) or "hard ROI" (e.g., measurable financial returns or savings). Table 4.1 outlines several examples of each. One hard ROI indicator that merits additional focus is cost avoidance or deferral. The example shown in Figure 4.5 outlines two separate paths for addressing an issue with associated costs for each.

Although regulatory bodies currently are not auditing medical device security programs within HDOs, the industry appears poised to head in that direction. The best way to justify a medical device security program is to focus on the hard ROI numbers and to evaluate HTM and facilities projects that involve upgrading or refreshing systems due to security concerns. In many cases, low-cost security controls can be implemented, thereby allowing these projects to be pushed out until the next refresh cycle with little risk. The costs can quickly add up and easily justify the investment into a medical device security program.

Mobile Medical Devices – Unique Considerations

Mobile medical devices have unique requirements when it comes to security. They have to meet all of the traditional wired medical device security best practices in addition to Radiofrequency requirements. Managing wireless

Table 4.1 Examples of Soft and Hard ROI Indicators

ROI Type	*ROI Description*	*Example*
Soft	Improved patient care	It goes without saying that having medical devices available at the right time and in a condition that is safe and effective for patient use is paramount for improving patient care.
Soft	Improved patient privacy	Scrutinizing and tracking protected health information use on medical devices and ensuring that patient privacy is addressed appropriately is advantageous for patients and reduces the risk of privacy breaches.
Soft	Institutional reputation	Being featured in a prominent news report for harming a patient due to a compromised medical device can reduce the likelihood of patients coming to the health system to receive care.
Soft	Patient retention or attraction	Institutional reputation is directly correlated to retaining existing patients and attracting new ones.
Soft	Predictable medical device performance	Predictable medical device performance is directly linked to patient safety, which in turn can affect institutional reputation.
Hard	Device utilization/ support contracts	Many medical device security management tools can also provide insight on device utilization and trend it over time. Scrutinizing utilization is a great way to renegotiate existing support contracts and curb spending on additional medical devices if the existing inventory is not being fully utilized. One example could be MRI utilization at different sites and spreading patients out to various facilities to make the most use of the equipment, rather than purchasing new MRI machines (which can cost upwards of $500K per machine).
Hard	Financial penalties/ HIPAA violation	If a HIPAA violation is traced back to a medical device, this can lead to financial penalties, as well as to scrutiny for not having a medical device program.

(*continued*)

Table 4.1 (Continued) Examples of Soft and Hard ROI Indicators

ROI Type	*ROI Description*	*Example*
Hard	Cost avoidance or deferral	The most powerful financial justification for a medical device security program is an HDO gaining the ability to discover, quantify, and prioritize high-likelihood, high-impact vulnerabilities. This will support the HDO in identifying cost-effective options that can be implemented quickly to reduce potential impact on patient safety. It is not always necessary to replace vulnerable medical equipment. In many cases, inexpensive technical, administrative, or physical security controls can be implemented to mitigate the risks and allow the HDO to continue using a medical device until they are ready to refresh it.

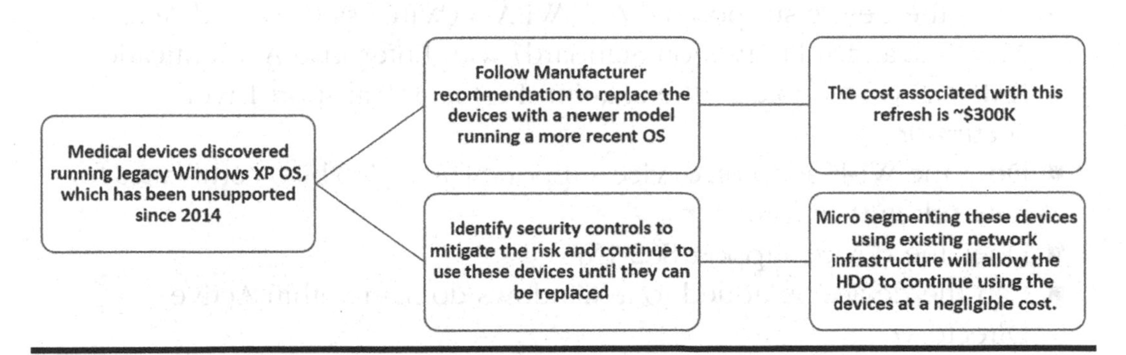

Figure 4.5 ROI indicators.

networks and RF is a highly specialized skillset and is often one that is lacking within MDMs and HDOs. Poor RF chipset or Infrastructure design can make the difference between a usable and an unusable medical device. RF interference can render a device unusable or unsafe. The emergence of Bluetooth use cases in healthcare extenuates this situation since the sensors relying on the 2.4 GHz ISM spectrum are susceptible to interference and can cause havoc on clinical and even patient experience workflows. Some of the questions that need to be posed when it comes to onboarding Wireless medical devices include

- the wireless capable medical device designed to be mobile, or stationary?
- Can the MTU size be manually modified on the device if needed?
- Can the device firmware be updated as wireless authentication and encryption mechanisms evolve in the industry?
- Does the device operate in the unlicensed RF Spectrum?
  - Is it IEEE 802.11a/b/g/n/ac/ax or any subset thereof compliant?
  - If not, what RF frequencies does it utilize?
  - Any restrictions on DFS or 802.11h channel announcements?
  - Can the device use a hidden SSID?
  - Any restriction with channel bonding?
  - Is it Wi-Fi certified?
  - What PHY rates are supported?
  - Can the device be set to a specific frequency band?
  - Is the wireless capability provided by a bolt-on bridge or an integrated wireless card?
  - What model of wireless card and chipset are used?
  - Is the device IEEE 802.11i compliant?
- Does the device support WPA-2, WPA-3 (Wireless Protected Access) AES (Advanced Encryption Standard) with Enterprise Authentication EAP-TLS (Extensible Authentication Protocol/Transport Layer Security)?
- Does the Wi-Fi adaptor/device support SHA-2 (256 bits) certificates for network authentication?
- Does the device support 802.1X?
- Can the device be added to a Windows domain within Active Directory?
  - Is the device IEEE 802.11e compliant?
- Does the device support WMM and/or WMM PS Mode?
- What queue is recommended?
  - Does the Wi-Fi adaptor/device support 802.11k (Neighbor List)?
  - Is the device IEEE 802.11r compliant?
- Is fast secure roaming supported?
- Is Opportunistic Key Caching supported?
  - Does the device support 802.11k?
  - Does the device support 802.11v?

Chapter 5

Wearable Devices and Remote Patient Monitoring

Walt Maclay

Use Cases in Healthcare

Patches for Glucose Monitoring

There are now patches that monitor glucose continuously. They are called Continuous Glucose Monitors or CGM. They are small enough to not interfere with daily activities, and they can be worn in the shower. These devices are more comfortable and less intrusive than collecting blood from the finger several times a day. Many young people don't want their friends to see them doing a glucose test, because it makes them different making a CGM device attractive. In the future, calibration may not be required making it even more convenient. They have a single needle that penetrates the skin to sample blood, although they are comfortable and can be worn for up to two weeks. It is necessary to calibrate the device when it is first installed. The patient pricks a finger and gets a reading with a standard glucose measurement device which is entered into the CGM device typically using a smartphone. Two leading players in this market are Abbott Diabetes and Dexcom.

There is a glucose monitor that fits into the eye. It has appeared in many reports, but it is only a laboratory study device. It has not been made into a commercial device. The accuracy has not been reported. It is unlikely that it meets FDA requirements for accuracy.

DOI: 10.4324/9781032690315-5

Insulin Pumps

Wearable insulin pumps have existed for many years. People in the diabetes community have hacked them and the CGM, creating a complete closed-loop wearable pancreas replacement. There is no commercial device available at this time, although work is being done to create one. Notably, manufacturers of insulin pumps and CGM have reduced their efforts to make their devices resistant to hacking. There seems to be an acceptance by these companies and the FDA that it is OK for hackers to create their own devices, as long as they don't sell them.

Patches on the Chest for Cardiac Monitoring

In recent years, a number of companies have developed patches that adhere to the chest for cardiac monitoring. The patches can be worn for a week at a time even in the shower. This gives a more complete picture of a person's cardiac condition, particularly for events that don't happen regularly, such as atrial fibrillation. They can collect and transmit data wirelessly. They are replacing Holter monitors for collection of cardiac data, because they are more convenient, lower cost, and can provide several days of data. It is common to send data wirelessly allowing these devices to be used for real-time monitoring. What is holding back adoption is the lack of AI software to monitor the data and detect anomalies that should be reviewed by healthcare professionals. With continuous monitoring, there is more data than anyone can effectively review. This limits the usefulness of real-time monitoring.

Cardiac monitors record ECG (electrocardiogram), but they may also monitor respiration, temperature, and blood oxygen.

Watches for Cardiac Monitoring

According to HealthTech, smartwatches are now helping healthcare providers collect and analyze a wider swath of data from patients between their appointments or after surgery. This data provides crucial and very valuable insights that can help identify possible and proper treatment.

The smartwatch trend, which has continuously growing sales every year, has inspired organizations such as Ochsner Health System in New Orleans. In 2015, Ochsner launched a pioneer program to better track uncontrolled hypertension among its patients. Stanford University's study in 2019 revealed that the Apple Watch could identify heart rhythm irregularities, such as atrial fibrillation, a leading stroke risk, which can be detected with 84% accuracy [1]. And to utilize this innovation, Ochsner now also utilizes the Apple Watch, which has benefited the doctors. The smart wearable device will send alerts about a patient's declining condition and send data to the healthcare in-charge's wrist, even if they are wearing gloves.

Samsung Electronics Co. Ltd launched the Galaxy Watch3 in August 2020. Galaxy Watch3 features a PPG sensor to monitor SpO2 levels. Also, with its enhanced accelerometer, the Watch3 smartwatch automatically detects hard falls. This smartwatch also records REM cycles, deep sleep, and total sleep time to score and help improve the quality of sleep.

These sensors are becoming popular in other wearable devices. Although watches are intended for personal health and not as medical devices, the Apple Watch did get FDA approval for detecting Atrial Fibrillation.

Implanted Cardiac Monitors

Implanted devices are a separate class of wearable device. They require a medical procedure to install them, but they can provide better data, because they can be placed where they work best. Implanted pacemakers and defibrillators have existed for many years. More recently tiny cardiac monitors have been developed. Medtronic makes the Reveal LINQ. It is only 1.2 cubic centimeters in volume, and its battery lasts for three years. At that time, it needs to be removed. It has a Bluetooth wireless connection. It is intended to monitor patients for various heart conditions that do not show symptoms for long periods of time. For many disease states, however, an external monitor can be used, reducing the risk and cost.

In Hospital Monitoring

Instead of having a nurse collect vital health data, hospitals are beginning to use a patch that is applied when the patient checks in. The patches can collect data much more often and at a lower cost. The data can be sent wirelessly to the patient's health record. These monitors can measure temperature, heart rate, ECG, breathing rate, blood oxygen, and blood pressure. As is described below, blood pressure measurement without a cuff is only just becoming available, so it is currently not generally used for patient monitoring. There is work being done to have artificial intelligence (AI) software evaluate the data and predict adverse events, sending help to the person-hours before the event can happen.

Sleep Monitors

Sleep monitoring is a rapidly growing area for wearable devices. There are many consumer products that monitor sleep, but there are also medical devices. A medical sleep monitor is typically a patch that adheres to the abdomen. Wearable devices can monitor sleep better than going to a sleep lab where it may be difficult for a person to sleep normally in an unusual environment. Wearable patches normally are wireless, so the person's movement is unconstrained. A wearable sleep monitor can be worn for several nights and collect more complete sleep information compared to one night at a sleep lab, and the cost is lower than a sleep lab. They typically monitor ECG, breathing rate, SpO_2, motion, and temperature. Sometimes skin impedance, also called galvanic skin response or GSR, is also measured.

Hearables

There has been development recently around hearables – devices in the ear for health monitoring. The ear is an excellent location for detecting heart rate, SpO_2, and motion. With millions of people already wearing earbuds to listen to music, it seems natural to add sensors to the earbuds. This has not taken off, perhaps because the elderly who most need medical attention are less likely to want to use earbuds.

Stimulation to Treat Disease or Pain

TENS or Transcutaneous Electrical Nerve Stimulation has been used for years to treat pain, such as back pain. An electrical signal is applied between two electrodes. Certain frequencies have been found to have therapeutic value.

More recently other wearable stimulation devices have appeared. There are devices that are implanted in the spine with an external control to treat intractable pain. This has become a major medical device area with several large competitors.

There is now a device worn on the wrist that treats essential tremors. Made by Cala Health the device is turned on and provides stimulation that is noticeable but not painful for approximately 45 minutes. Relief is said to last for many hours after treatment.

EEG Monitoring

EEG (electroencephalogram) monitoring has been changing. New devices are getting useful information without a large number of leads placed all over the head and requiring shaving of hair. Ceribell makes a head band for monitoring the brain during a seizer. It has 10 electrodes around the band. No hair needs to be shaved. The EEG data is run through an algorithm to identify if the patient is having a seizure.

Magnetic Field on Head to Stop Migraines

Magnetic fields are being used to stop or reduce migraines with a device that mounts on the head. Due to the power consumption, these devices may need to plug into power or use large batteries, so they may not be considered wearable devices.

Augmented Reality and Virtual Reality Glasses

Augmented reality glasses allow you to see your surroundings. The added content appears in front of the real world. You can get instructions or access patient records without touching anything. When fully implemented, this could really help in surgery and ER wards.

Virtual reality glasses display an entirely virtual world, and the wearer does not see the real world at all. This is primarily useful for training, where you may have a virtual hospital and virtual patients. A computer can control the learning and pace of learning.

Training of healthcare professionals is becoming more important as technology keeps changing the way they perform their jobs. Some training is best done without a live patient. Other training requires a live patient. Both are benefiting from the use of augmented reality glasses and virtual reality glasses.

Physiological Sensors Used in Wearable Devices

Body Temperature Measurement

Temperature sensors are low cost and there are many types available. Measuring core body temperature, which is usually desired, is not simple, however. Skin temperature is often lower than the core body temperature, especially at the extremities. Many wearable devices are on the wrist, which is not a good place to make this measurement. The forehead, under the arms, and in the ears are good places, but most wearable devices are located elsewhere.

Hearables, wearable devices in the ears, now exist. Many people wear earbuds for long periods to listen to music. Sensors can be placed in these devices without making them much bigger. Besides temperature, the ear is a good place to measure heart rate and blood oxygen.

Software can combine data from multiple sensors, "sensor fusion," to determine when skin temperature is likely to be near the core body temperature. If the skin is wet, the person may be perspiring or in a shower. In either case, the skin temperature is not a good indicator of core body temperature. If the person has been outside in cold weather, the skin is likely to be cold. If the room temperature is moderate and the person has been moderately active, the skin is likely to be close to the core body temperature.

Wet skin is likely to be colder than the core body temperature. Moisture on the skin can be measured using GSR. The electrical impedance of the skin is measured with electrodes. The ambient temperature can be measured with a temperature sensor that is kept away from the skin. Activity can be measured with a motion sensor. These sensors can be used by software to do sensor fusion.

Motion Measurement

Motion of the body has been measured for decades. Step counters were originally mechanical devices used to estimate the distance walked or run. Now low-cost motion sensors have replaced the mechanical devices. They are very small and consume very little battery power, so they are used in many devices as auxiliary sensors, sometimes for sensor fusion.

Step counting is a good measure of activity and is used in many consumer devices. The manufacturers of the motion sensors have developed advanced software that is able to measure step counts when mounted on the wrist or other places on the body. This is quite a feat, although it is not perfectly accurate. The software can work with the motion sensor on the wrist, ankle, or torso. The software algorithms are even able to determine with reasonable accuracy that a person is walking, standing, or sitting.

Motion sensors are also used to measure motion during sleep. With software interpreting the data, it is possible to measure the stage of sleep with good accuracy. This is important for sleep analysis as well as for consumer products.

Motion sensors can measure gait which can be used to indicate several conditions, such as dementia and Parkinson's disease. The specificity of the indication is only moderate, but it is good enough to refer people for further diagnosis by a healthcare professional.

Another use is for dead reckoning – tracking someone's motion. Motion sensors are only accurate for a few minutes. They accumulate errors over time, but because of their low power, they can be used as a substitute for GPS, which is moderately power hungry. The GPS can be turned on only every few minutes to save battery power, and the motion sensor can track the position while the GPS is off.

Heart Rate Measurement

There are several ways to measure heart rate. ECG electrodes may be used. Two electrodes located on most parts of the body can pick up a good enough signal to measure heart rate, even where the signal is not sufficient for an ECG measurement. A pulse plethysmograph (PPG) sensor can be used to measure heart rate. It was originally used to measure blood oxygen, but the heart rate is a stronger signal that needs to be removed to sense oxygen. These sensors work quite well even on the wrist.

It is also possible, but infrequently done, to measure heart rate with a pressure sensor located over an artery. The pressure pulse can be sensed just as a person can feel the heart rate by placing fingers on the inside of the wrist.

For either an ECG electrode or a PPG sensor, it is important that there be good contact with the skin. On a wrist device, this may be uncomfortable, presenting a design challenge. In many cases, heart rate is not needed continuously, and software can determine when the signal is good. When it is not good the heart rate can be ignored.

Blood Oxygen Measurement

Oxygen saturation or SpO2 is only measured with a PPG sensor. A pulse oximeter uses a PPG sensor. Originally, they were clipped onto a finger. Light of at least two wavelengths is passed through the finger. Both wavelengths are sensitive to the pulse. One wavelength is absorbed by hemoglobin. One signal is subtracted from the signal of the other wavelength to remove the pulse. The result is an accurate measurement of SpO2. This is a transmissive pulse oximeter where the light passes through the body.

The transmissive PPG measurement is only possible on the finger or ear, where light can pass through. For other locations, a reflective PPG measurement is used. The measurement uses similar wavelengths of light and subtracts the pulse from the signal. The reflected signal is much weaker than the transmitted signal, so the measurement is more difficult. More signal processing is required. It does not work where the body does not have good blood perfusion, but good results have been achieved on the wrist.

ECG, EEG, and EMG Measurement

All of these signals are voltages generated by the body. The sensor consists of electrodes to pick up the voltage and an amplifier to measure the tiny signal.

ECG has been measured for a long time with laboratory equipment that typically uses 12 leads and wires to the equipment. Signal processing has improved to the point that single-lead ECG measurements are almost as good as 12-lead measurements. Wearable devices almost always use two contacts (which is called single lead). The contacts are usually dry for convenience, although it is easier to get a good signal with wet electrodes. The contacts need to be spaced at least 4 cm apart to get a good signal. ECG cannot be measured on the legs or arms. On the head, it would be obscured by the EEG signal. Successful measurements have been made in pants where the electrodes are on the lower abdomen. There are implanted ECG monitors, but they are not as widely used as non-invasive wearable devices.

EEG has been measured with electrodes placed on shaved areas of the scalp with wires to the equipment. Many non-critical EEG applications, such as for consumer products, use only two electrodes. The electrodes can be on the temples where the electrodes may be attached to glasses or a head band. Successful measurements have been made with a helmet that has electrodes at the end of projectiles that reach the scalp without shaving any hair. This is critical, as a wearable device that requires preparation, such as shaving, is inconvenient. The measurement can only be done on the head. The signal elsewhere is too small.

EMG (electromyogram) is the measurement of the signal that activates muscles. It is not a common medical test. It can be used to sense what muscles are moving, or in the case of people with stroke, the muscles that the brain has directed to move, even if they did not move. The electrodes need to be carefully placed. On the forearm, for example, you can sense the individual muscles moving the fingers, but they are only a few millimeters apart, and the arm does not have a good reference for accurately positioning electrodes.

Respiration Rate Measurement

The number of breaths per minute can be measured with several techniques. An old and still viable way is with a chest strap. A sensor measures the change in length of the flexible strap as the chest moves. This is fine for a shirt with sensors. Most wearable devices are small or not located on the chest, and this technique is not suitable in those cases.

A nasal cannula can be used in a hospital. It is not convenient or comfortable for a wearable device.

Thoracic impedance is an accurate technique. The electrical impedance of the chest varies as the chest expands and contracts. A sensor similar to the GSR sensor measures the impedance. It is important to measure deep in the tissue and not the skin at the surface so that changes in skin conductivity do not interfere. This sensor works well on the chest, but it has not worked on the wrist, although efforts have been made on this. The arm, being narrower than the chest, has a much higher impedance. Thus, the overall impedance is dominated by the arm, and the signal is too small to measure accurately.

Respiration rate can also be measured directly from the ECG signal after filtering out the heart rate. As long as the breathing rate is well below the heart rate, this can work, as the heart is affected by the movement of the chest during breathing.

Blood Pressure Measurement

Up until 2021, the only way to get medically accurate blood pressure measurements without calibration on each person was with a cuff. This has changed. It is now possible to do this with at least two technologies.

A PPG sensor with advanced software can measure blood pressure. This was a big challenge, partly because the measurement is sensitive to skin color and motion. The different wavelengths of light pass through the skin differently. A clinical study was done by Valencell, Inc., that achieved medically accurate blood pressure measurement without calibration [2]. Now a small wearable device can measure blood pressure. The study used a sensor on the finger. They are working to achieve this result on other parts of the body.

Another technique was just disclosed by PyrAmes Inc. They detect the pressure of the pulse passing through the arteries. The pressure causes movement which is detected by a capacitive sensor [3]. Software needs to not only calculate the pressure but also determine when the signal is not accurate. This was not a clinical study, so it still needs work before being used in a medical device.

Another technique, Pulse Transit Time, has been used but has not achieved medical accuracy without calibration. It measures the time difference between the beat of the heart (measured with ECG electrodes) and the arrival of the pulse at an extremity (often measured with a PPG sensor). The time difference is proportional to the blood pressure. Although it is not medically accurate, it can be used in applications where the *change* in blood pressure is important, and the absolute value is not. Detecting a change is often important to indicate the need to take measures, such as visiting a healthcare provider.

Blood pressure has also been measured in a laboratory setting using the ECG signal and advanced neural network software. It is not medically accurate, however.

Blood Glucose Measurement

The standard for blood glucose measurement is a sample of blood from the finger placed into a sensor. A very large amount of money has been expended to find a non-invasive glucose measurement technique. Today this has been achieved, although not fully, with CGM.

Abbott Diabetes has a small patch that is worn on the arm held in place with adhesive. It has a needle that passes through the skin to reach the blood, so it is technically invasive, but it is comfortable to wear for more than a week at a time. It requires calibration with a finger prick each time the disposable device is replaced. This is much more convenient than pricking your finger several times a day, however.

Another technique is a patch with microneedles. The microneedles only pass through the outer layer of the skin. They don't look or feel like needles. The texture is similar to sandpaper. It is challenging to get good results without intimate contact with the blood. This device also requires calibration.

Both of these sensors must be placed where there is good blood perfusion. They do not work on the wrist.

There are also implanted glucose monitors. Most people prefer a less invasive device, however.

Where the sensors can be located on the body

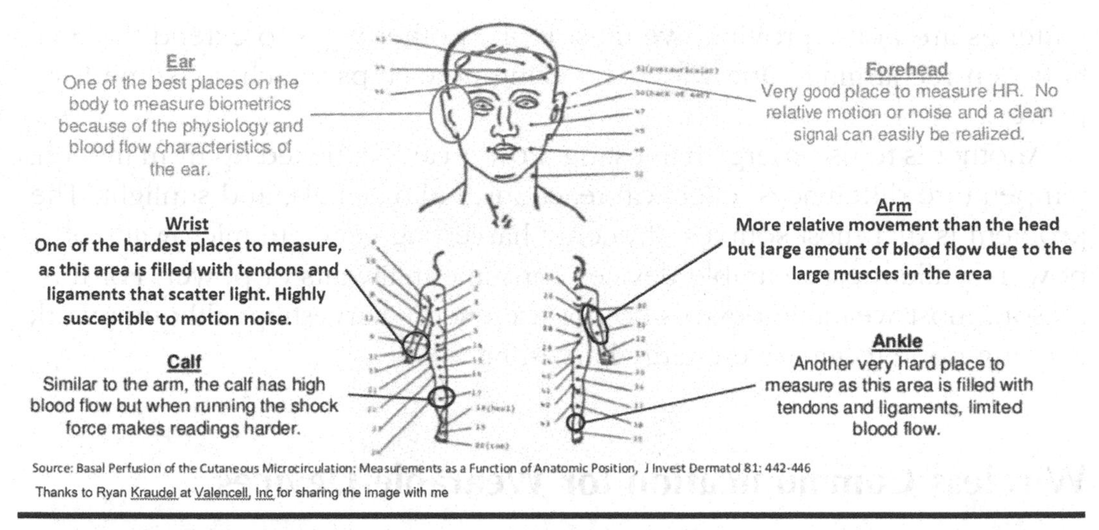

Figure 5.1 PPG sensor placement.

Many wearable devices are designed to go on the wrist for convenience, but that is a poor place for most measurements.

ECG measurement is generally only done on the torso. The chest is clearly the best place because of its proximity to the heart. Successful measurement in pants has been done using dry electrodes even when the person is moving. The Apple watch and other wrist-worn devices can only measure ECG when the opposite hand touches the watch, providing a signal that is from one arm to the other. This works well, but only while the watch is being touched.

PPG sensors for blood oxygen measurement require good blood perfusion. The wrist is challenging, but it has been used successfully. Getting a pulse measurement from a PPG sensor is easier and can work nearly anywhere on the body, as illustrated in Figure 5.1.

Temperature measurement is best done under the arm, on the forehead, or in the ear. In general, skin temperature is not reliable at the core body temperature, except at these locations. On the other hand, it can be useful to measure the temperature at extremities, not core body temperature, to detect problems with circulation and certain diseases.

Batteries

Nearly all wearable devices use batteries, and batteries are not improving very fast. They are approaching the limit of chemical energy density. Since batteries are not improving, we must look at other ways to extend the time between recharging. One way is by using new chips which consume less power.

Another is to use energy harvesting. Power can be picked up from motion, temperature differences, chemical reactions, radio signals, and sunlight. The problem is that most sources of energy harvesting generate microwatts of power, while most wearable devices consume milliwatts of power. For this reason, most wearable devices do not use energy harvesting, although work is being done to improve energy harvesting.

Wireless Communication for Wearable Devices

Nearly all wearable devices communicate the sensor data through a wireless connection. There are some devices that use a connector to download data, but that means the data is not available immediately. There are many choices for wireless communication. The most commonly used is Bluetooth Low Energy or BLE. It uses the lowest power, and it can communicate directly to most smartphones. It must be within a few feet of the phone, however, so it will not work unless you take your phone wherever you go.

WiFi is convenient because it is available on all smartphones, but it consumes a lot of power, which requires a large battery. Cellular service can connect directly to the Internet rather than pass through a phone. This is a big advantage when a phone is not always being carried. Cellular uses a very large amount of power, however.

There is a relatively new class of wireless called Narrow Band. This includes NB-IoT, LTE-M, LoRa, and Sigfox among others. Narrow Band communication transmits several kilometers at low power. The trade-off is a low data rate, but that is not a problem for most wearable devices. They rarely need to send more than a few hundred measurements per second, which is considered slow, and they often send less than one sample per second.

Figure 5.2 compares various wireless standards. The data rate increases from left to right. The communication distance increases from top to bottom, and the power is shown as a number in milliwatts in Figure 5.2.

One important wireless standard is not shown in this table: NFC or Near Field Communication. It is similar to RFID, which is often used as a substitute for bar codes. NFC requires the transmitter to be within a few centimeters of the wearable device, and it requires no power from the wearable. The device doing the reading provides wireless power. Many smartphones have NFC

	100 bps	10K bps	40K bps
1 m	BLE4/Zigbee 0.15 BLE Mesh 0.15 Bluetooth 25 WiFi 50 LoRa 0.5	BLE4/Zigbee 7.5 BLE Mesh 7.5 Bluetooth 25 WiFi 50 LoRa 10	Zigbee 30 Bluetooth 25 WiFi 50 LoRa 20
50 m	Zigbee 20 WiFi 100 LoRa 0.5 Sigfox 0.5 NB-IoT, LTE-M 1.0 LTE, 5G Cellular 100	Zigbee 30 WiFi 100 LoRa 20 NB-IoT, LTE-M 30 LTE, 5G Cellular 150	WiFi 200 NB-IoT, LTE-M 200 LTE, 5G Cellular 200
1 km	LoRa 30 Sigfox 30 NB-IoT, LTE-M 20 LTE, 5G Cellular 120	NB-IoT, LTE-M 100 LTE, 5G Cellular 200	NB-IoT, LTE-M 400 LTE, 5G Cellular 400

Figure 5.2 Power – how much, how far?

built in. They are often used to scan a device, such as a glucose monitor. The data is immediately visible on the phone, and it can be transmitted to the Internet to be shared.

Adhesives for Attachment

Since the torso is the best place for most measurements, there are many devices that attach to the torso. They usually use an adhesive patch. This is convenient, but there are limitations. A patch can only be applied to the skin for one or two weeks before the skin becomes irritated and begins to break down. For most applications, the adhesive needs to perform well while bathing or showering.

An alternative is to have sensors in a shirt. When the shirt is put on, the sensors begin recording and transmitting data. Shirts have limitations too. They must be washed, and it is challenging to have electronics survive a large number of times in a washing machine. Typically, devices are tested for 50 washes, but some clothing is washed more than that.

Sensors in a shirt need to be flexible, and ideally stretchable. Flexible circuits have been around for decades, but they need to be protected from moisture. Good sealing methods are more recent. Flexible circuits are barely out of the laboratory. They are not yet widely used in wearable devices.

References

[1] Stanford Medicine News Apple Heart Study demonstrates ability of wearable technology to detect atrial fibrillation, March 16, 2019.
[2] [https://valencell.com/featured/valencells-cuffless-calibration-free-blood-pressure-monitoring-technology-selected-to-present-at-american-college-of-cardiology-annual-scientific-session/]
[3] *Journal Sensors*, published by MDPI, Basel, Switzerland, June 2021.

Chapter 6

Explosion of Robotics in Healthcare

Roger Smith

Beginning

Given all of the potential entrees for robotic devices to enter the healthcare field, it is interesting that the first documented application was in surgery in 1983. The "Arthrobot," developed by Dr. James McEwen, Geof Auchinlek, and Dr. Brian Day at the University of British Columbia, was used to assist in the positioning of a patient's leg during an orthopedic surgery. Arthrobot could manipulate and hold the leg in an exact position to support the incisions and bone cutting that would be performed by the surgeon. A second more interventional case of robotic surgery assistance was the 1985 conversion of the PUMA 560 industrial robot to assist with a brain biopsy. The PUMA used its mechanical accuracy to insert a biopsy needle at a precise angle, to a specific depth, and extract a brain tissue sample backward along the same path achieving a sample location accuracy of 0.05 mm. These early cases opened the doors for an explosion of the technology that would expand to a multi-billion-dollar industry over the next three decades.

While the robotic surgery race has been very crowded and publicly visible, something similar has been occurring with equal fervor in multiple healthcare domains. Robotic systems can now be found in hundreds of different departments supporting all aspects of healthcare.

DOI: 10.4324/9781032690315-6

Technologies

Healthcare has always been a rich user of technology. Though it is often criticized for being slow to adopt a new technology when compared to manufacturing, defense, ecommerce, and social media industries, once it does embrace a technology like robots, it quickly becomes one of the largest users of that technology. Within any community, the local healthcare systems are among the largest users of technologies like computers, networks, databases, energy devices, and robots. In most cities, there are more robots in the healthcare system than there are in any other industry in town. Since healthcare is a local business at its core, these systems are duplicated from one city to the next.

If we examine the component technologies that make up the hundreds of robots that are currently servicing healthcare, we find essential pieces that have made significant advances in the last three decades. Robotic assistance in healthcare evolves by leveraging the following (Figure 6.1):

- Mechanical: Improving physical actions.
- Electronics: Extending human control of mechanical devices.
- Sensors: Extending, augmenting, and replacing human senses.
- Manipulators: Refining and regulating human movement.
- Connectivity: Global accessibility and control.
- Intelligence: Enhancing human thinking and communicating.
- Metaverse: Creating shared information spaces.

Different combinations of these technologies form all the robotic devices we are about to explore. Each application area has multiple competitors offering unique designs for devices that perform a specific service. This richness of solutions is a testament to the versatility of the technologies and the creativity of the designers and engineers who create them.

Applications

We have identified over 20 unique healthcare applications of robots that are currently in operation. These applications are aligned into a smaller number of categories based on the services that are rendered by the robots (Figure 6.2).

The use of robotics in all of these areas is much more prevalent that most healthcare administrators or clinicians realize because each is likely to

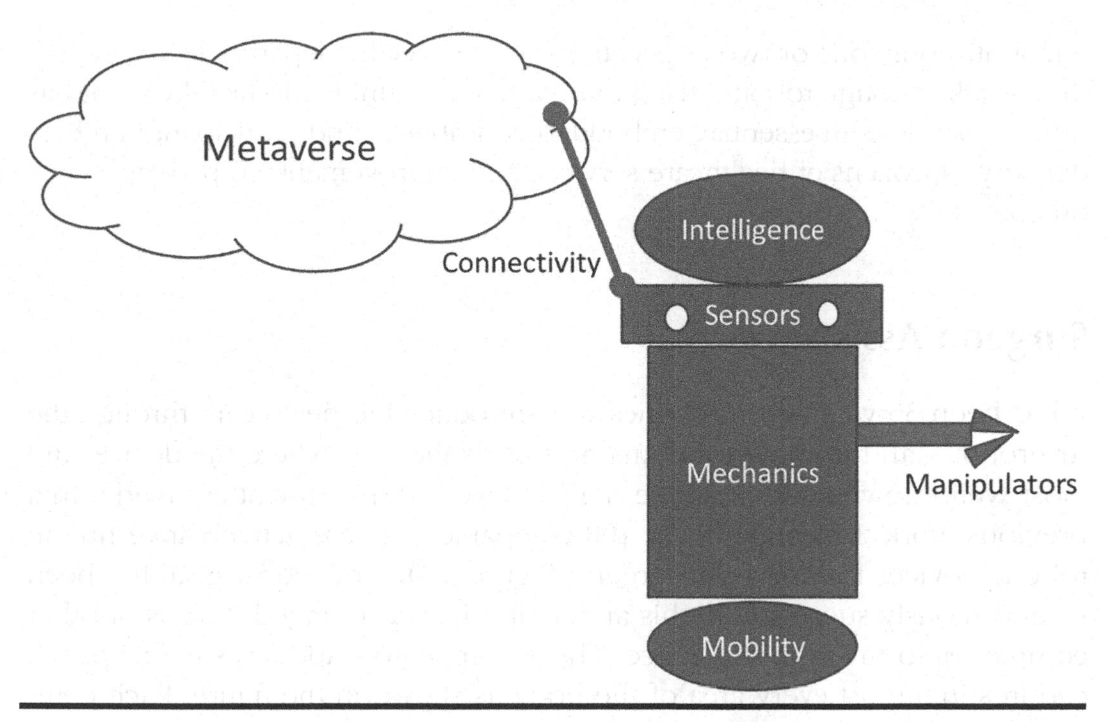

Figure 6.1 Technologies that contribute to healthcare robots that can make a positive contribution.

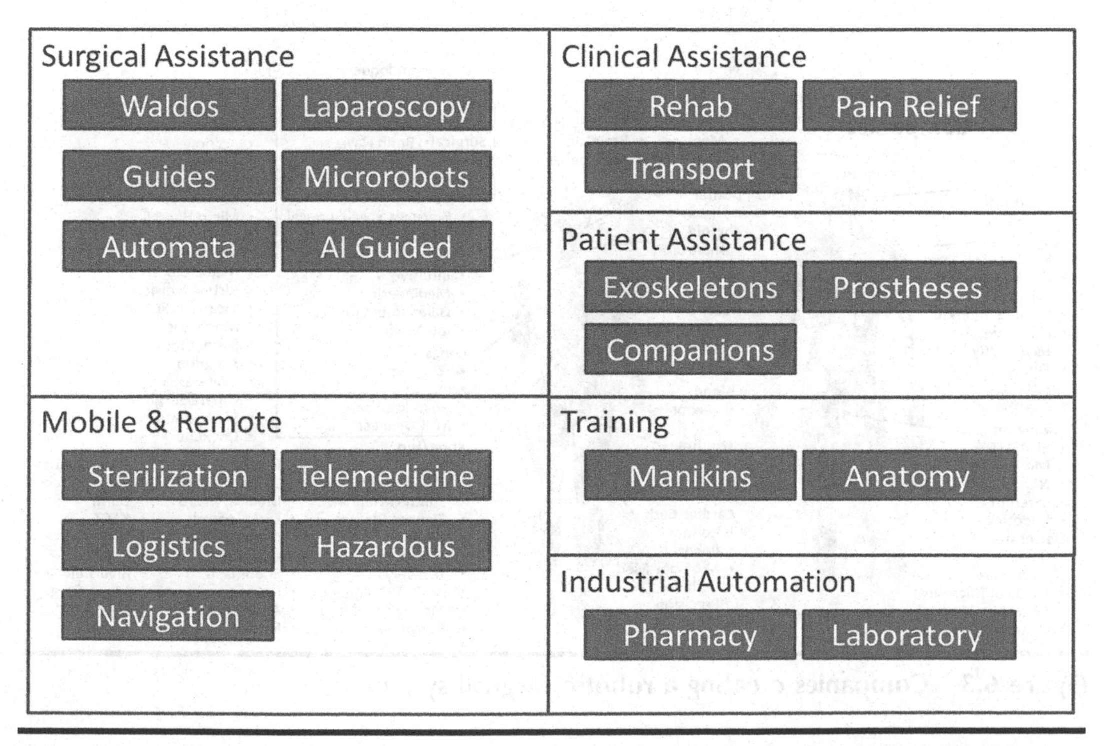

Figure 6.2 Six categories and over twenty unique applications of robotics in healthcare.

encounter only one or two applications in the specific department where they work. Though robotics used to be a rare and unique in healthcare, it has quickly become an essential, embedded contributor, and participant in the delivery of dozens of healthcare services and improvements to patient outcomes.

Surgeon Assistance

It has been 35 years since robotics was introduced to healthcare through the Arthrobot – and the surgical domain remains the area where the devices are most well-known and where the highest levels of investment are made. In a previous work, I identified over 100 companies that are actively investing in robotic devices to assist with surgery (Figure 6.3). Intuitive Surgical has been so enormously successful in this area that it has encouraged dozens of other companies to invest in this space. These competitors address surgical procedures in almost every area of the body as shown in the figure. Each company represents several million dollars of investment, measured for the size of this market area.

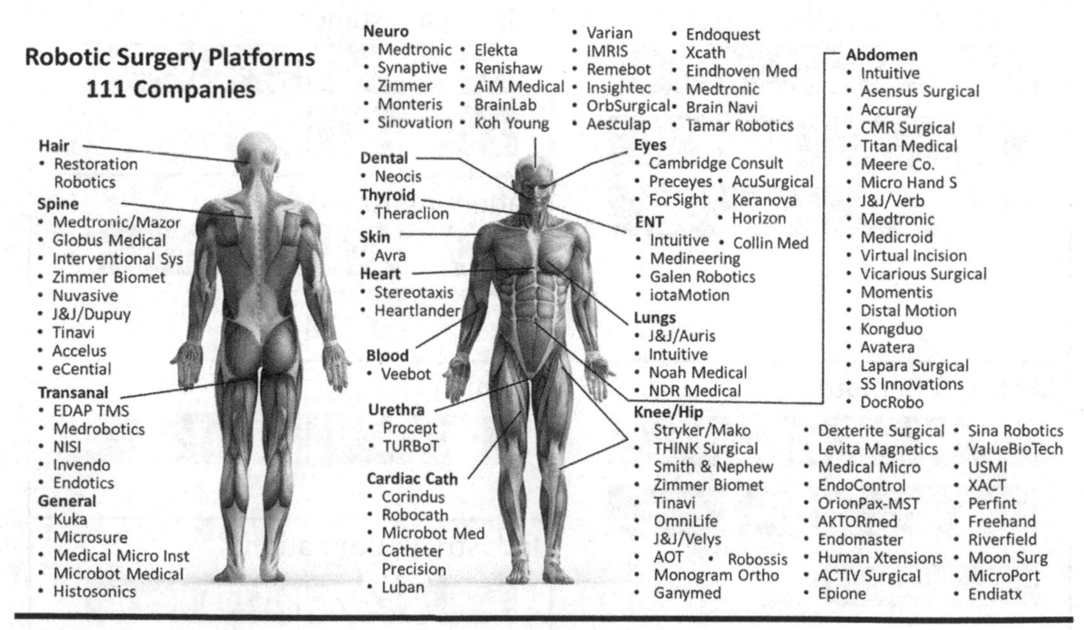

Figure 6.3 Companies creating a robotic surgical system.

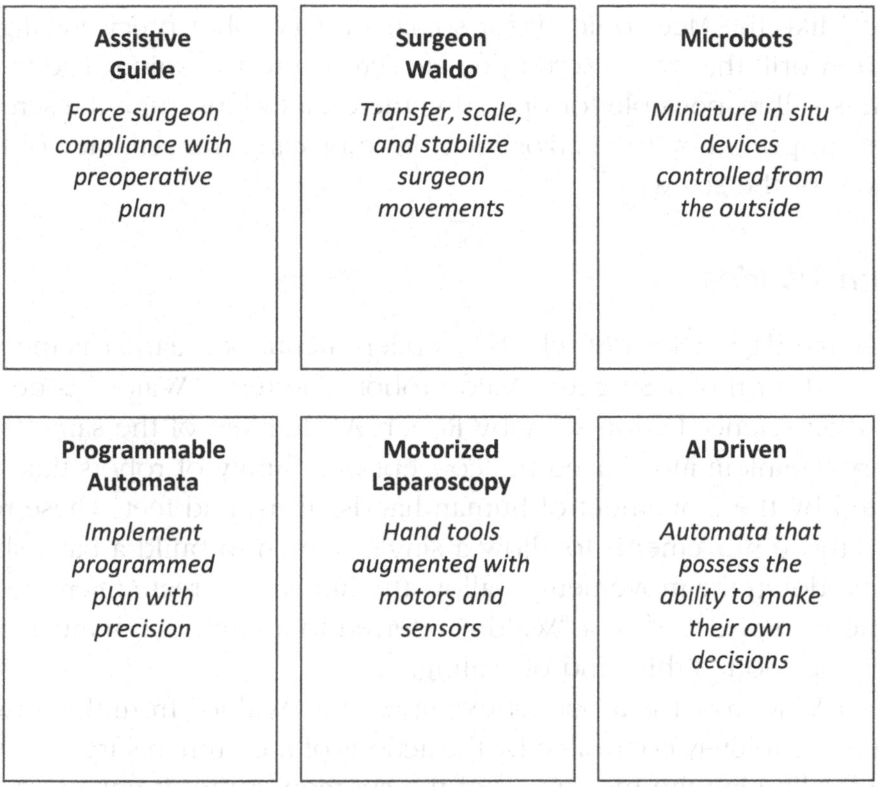

Assistive Guide
Force surgeon compliance with preoperative plan

Surgeon Waldo
Transfer, scale, and stabilize surgeon movements

Microbots
Miniature in situ devices controlled from the outside

Programmable Automata
Implement programmed plan with precision

Motorized Laparoscopy
Hand tools augmented with motors and sensors

AI Driven
Automata that possess the ability to make their own decisions

Figure 6.4 Six types of assistance for robotic surgery devices.

Examining each of these surgical devices, we have grouped them into one of six types of assistance provided to a surgeon (Figure 6.4). Four of these are well established with multiple products in each. The other two, microrobots and AI-driven robots, are just emerging from research laboratories with a few demonstrations of their capabilities.

Assistive Guides

Several robots have been created with the specific objective of reducing variation in procedures that originate from the physical movements of the human surgeon. Inserting metal screws into the spine and shaving bone from a knee or hip is a physical activity that requires a strong and steady hand from the human surgeon. In spite of their experience and skills, these human movements are fraught with variations that are not always intended or beneficial. A robotic device can regulate the movement of that hand and the instrument it is using to reduce the unintended variation by several orders of magnitude.

Robots like the Medtronic Mazor spinal surgery robot precisely align a guide for a drill that will insert a pedicle screw into the spine. The human surgeon is still responsible for operating the drill and inserting the screw, but they are supported by the Mazor robot in improving the accuracy of the placement of the screw.

Surgeon Waldos

The Intuitive da Vinci robot which has been mentioned earlier is the most widely used form of a Surgeon Waldo robot. The term "Waldo" is derived from a 1952 science fiction story by Robert A. Heinlein of the same title. In that story, Heinlein introduced the concept of a family of robots that are controlled by the movement of human hands, arms, and feet. These robots scale up those movements to allow a single human to build a battleship or they scale down the movement to allow the human to create microelectronic components. For decades, a "Waldo" referred to a family of manufacturing tools that supported this kind of scaling.

The da Vinci robot is a perfect example of a "Waldo" from that original story. It is completely controlled by the actions of the human surgeon. It is not a guide for limiting the movement of the surgeon, rather it enhances and magnifies the actions of the surgeon. It allows them to operate on tissue that is more difficult to reach with manual surgical tools. It uses magnification of vision and scaling of hand to allow them to operate at a much smaller scale and with more precision than they could without the aid of the robot.

Programmable Automata

Robotic radiotherapy offers improved precision in delivering energy at a specific location, in specific amounts, and for a specific time. These devices are programmed with the geometry of the patient, the tumor, and the physics of energy propagation. Using this information, they can function autonomously to precisely apply the right amount of energy to specific locations in the body. In fact, the human manual performance of these tasks is so fraught with errors and variation that it is often unsafe.

The CyberKnife system is a non-invasive treatment for cancerous and non-cancerous tumors and other conditions where radiation therapy is indicated. It is used to treat conditions throughout the body, including the prostate, lung, brain, spine, head and neck, liver, pancreas, and kidney, and can be an alternative to surgery or an option when the patient's condition is inoperable.

CyberKnife is an image-guided linear accelerator that was specifically designed to deliver stereotactic radiosurgery (SRS) and stereotactic body radiation therapy (SBRT). It is the precision of the system, delivered by its robotic arm, and real-time adaptive delivery of the radiation beam to the tumor throughout treatment, that makes a difference for patients. CyberKnife's accuracy is at sub-millimeter levels, which can significantly reduce the risk of side effects from over exposure to healthy tissue or under exposure to cancerous tumor tissue.

The robot moves and bends around the patient, to deliver radiation doses from potentially thousands of unique beam angles, significantly expanding the possible positions to concentrate radiation on the tumor while minimizing dosage to surrounding healthy tissue.

Motorized Laparoscopy

Laparoscopic surgery was invented in the 1960s and became very widespread in the 1980s. It required the creation of a number of new instruments with long shafts, allowing the surgeon's hands and instrument controls to remain outside the body, while the grasper, scissor, or energy tip entered the body through a small incision. As the mechanical and electronic components of the large robots in the above categories became smaller and cheaper, it enabled the addition of some of those robotic features to individual hand tools used in laparoscopy. Dozens of manufacturers created these enhanced, motorized laparoscopic tools for use in various procedures, generally in the abdomen.

Human Xtensions Ltd. has created the HandX robotic instrument. The motors in the handle and the unique design of the gripper allow the surgeon to control a grasper or energy hook at the end of the tool with six degrees of freedom (6 DOF). This full-wristed motion at the end of the instrument is not possible with traditional laparoscopic tools, and it is one of the most popular features of large robots like the da Vinci.

Microbots

Microrobots have shown significant potential to conduct microscale tasks such as drug delivery, cell manipulation, micro-assembly, and biosensing using manual control. The unique challenge with these miniature devices is achieving traction to enable mobility. The robots are so small that their own mass is not sufficient to leverage gravity and friction as the enablers of mobility. Multiple approaches have been taken to create a substitute, including applying external magnetic fields, equipping them with tiny grippers, attaching hooks on the tires and feet, and leveraging the flow of blood or digestive actions to move the robot.

In one case, the application of targeted delivery was accomplished using magneto-tactic bacteria under DC magnetic fields. Other research groups have also explored microrobots for transporting target objects such as cells and chemicals using magnetic fields. Microgrippers have been developed for microrobots using micro-electro-mechanical systems, a technology that can be used to improve the functionality of microrobots. Researchers created tiny devices that can deliver drugs to the body by attaching themselves to a person's intestines.

In a gastroenterology experiment at Johns Hopkins University, the scientists took inspiration from a hookworm creating shape-shifting microdevices called "thera-grippers" that can mimic the worm and latch on to the intestinal mucosa of a patient. The six-pointed devices, each as large as a dust speck, are made of metal and thin film that can allow them to change shapes. They are covered by a heat-sensitive paraffin wax and have the potential to release a drug gradually into the body. The scientists say that thousands of such devices can be let loose in a gastrointestinal tract. As the wax coating on the tiny robots matches the body's temperature, thera-grippers automatically close and latch onto the wall of the colon. As they attach and dig into the mucosa, they start slowly releasing the stored medicine. In time, the devices lose their grip on the intestine tissue and leave the organ through normal gastrointestinal function.

AI Guided

Each of the above devices is guided or programmed directly by a human surgeon, clinician, or multi-specialty team. As the understanding and encoding of each procedure becomes more complete it will be possible to provide instructions to AI software that can accurately perform the function and independently make decisions based on sensor data collected. AI-guided robots will remain under the supervision of a human clinician but may be able to demonstrate that they can perform a routine procedure at a level equal to the human clinician. Once encoded, collecting data, and updating its model, an AI algorithm should be able to improve its performance more quickly than a human and share that knowledge with every other similar robot on the planet. This level of education and dissemination is impossible for globally distributed human practitioners who distribute their expertise via journal articles and conference presentations.

Clinician Assistance

Outside of the surgical suite, robotic devices are helping members of the clinical team deliver their services in hospitals and clinics. These devices usually allow a human to focus on activities that are at the top of their certifications, perform tasks with more precision, or deliver services more efficiently.

Rehabilitation Robots

Rehabilitation from injuries has always posed unique challenges. Cerebral and motor stimulation are both required while using the limb under rehabilitation. Simple movement patterns and even passive exercise routines do not lead to maximum recovery of a limb's capabilities. Rehabilitation supported by robotic systems has numerous advantages: (1) it allows more intensive and tailored rehabilitation activities and services, thus increasing the amount and quality of therapy that can be administered; (2) it allows all the involved actors on the clinical team (e.g., physiotherapists, physicians, bioengineers and other figures) to set and manage work parameters to make the rehabilitation specific and optimal for the patient, this includes the type of exercise, the level of assistance from the robot, the force, and kinematics that the patient must exert; and (3) the computers can collect objective data on performance and use that to measure progress over time.

Robots, especially when coupled with imaginative and graphic computer exercise programs and games, have higher levels of flexibility and variability in creating engaging programs that encourage the therapist and the patient to be more involved with the program both cognitively and physically.

Devices like Barret Medical's BURT and TyroMotion's DIEGO robots have combined robotic-supported exercises with engaging gaming environments that challenge the patient to achieve game objectives while performing the prescribed exercises. Engagement with the game can increase the time that a patient is willing to spend on rehabilitation and their level of cognitive engagement during the tasks. Both of these make significant contributions to the recovery of functionality. The self-guided play of the games also reduces the demands on the human physical therapist, allowing them to provide services to more patients simultaneously.

Pain Relief

Service robots with human behavioral sensing for clinical or personal use in the home have attracted a lot of attention thanks to their advantages in relieving high labor costs of human assistance. One instance of this is the emergence of a robotic device that can provide custom-programmed massage therapy. The hope is that the availability and anonymity of these devices will attract patients who are otherwise hesitant to engage in the more intimate traditional massage environment. Robotic devices also offer the advantage of programs that are customized for a specific patient, exactly repeatable, and with a potential for measurable progress toward physical and healthcare goals.

Capsix Robotics in Lyon France was founded with the aim of offering this type of robotic massage. Together with Kuka, the German robotic giant, and LBR Med, they have created the iYU robot, which can almost entirely replicate the work of a massage therapist. The core hardware components are provided by Kuka. LBR Med then customizes the system for massage applications, adhering to the necessary safety, sanitation, and compliance regulations of the healthcare field. Capsix then sells and services the devices for client locations.

Massage Robotics Inc. has launched a larger luxury robot that is equipped with multiple attachments, is controlled by neural networks created in collaboration with Google, and communicates what it learns with other installations of itself.

Pain management is also being addressed psychologically by Canadian researchers who have utilized a robotic device to effectively reduce pediatric cancer patients' distress during needle insertion. The humanoid MEDIport robot is placed next to the child at eye level and performs a pre-programmed series of behaviors, like as dancing, to distract the patient before, during, and after the needle insertion. The robot was found to significantly reduce distress.

Transport Patients

Patient transport is one of the essential, but menial, entry-level positions in a hospital system. It requires no unique education so provides an employment entrée for almost anyone. Most hospitals require that a new employee in patient transport remain in the department for 6–12 months. But, at the completion of this period, the transfer rate out of transportation and into other departments is 100%. Therefore, the department is constantly losing staff who know the hospital facility well enough to deliver patients reliably. An intelligent wheelchair that has been programmed to navigate the facility flawlessly would be an invaluable asset. It would be able to identify the fastest route, find alternatives when congestion or construction occurs, navigate around people and equipment in hallways, and anticipate the queues that are forming at the destination.

Multiple universities have created prototype robotic wheelchairs with these capabilities. Northwestern University is developing a shared-autonomy wheelchair that is customized to the physical needs and personal preferences of the passenger. It also focuses on simple integration with existing chairs and control interfaces to mitigate the costs and leverage the insurance coverage for these existing wheelchairs. They have developed perception algorithms that can detect doorways, inclines, drop-offs, and docking locations.

Robotic wheelchairs could empower people with physical limitations in navigating their own homes without extreme physical exertion or the dexterity to operate the traditional controls of a motorized wheelchair. Large public spaces, including malls, big box stores, and government facilities, could empower guests to navigate their spaces with confidence. These devices would travel the best route to any store, product, or department without requiring guests to interpret "You Are Here" maps or wander uncertainly around the halls.

Patient Assistance

Robotic devices are not limited to clinical facilities, they can also accompany a patient home and assist them in living with conditions that previously would have severely limited their lifestyles.

Exoskeletons

Every year, nearly 56 million people suffer from acquired brain injury, 15 million suffer from stroke, up to 500,000 people suffer from a spinal cord injury, and 2.8 million people live with multiple sclerosis. Many of these people are left with limited mobility or some form of paralysis. This can be a devastating diagnosis that is completely life-changing for both patients and their families. For those who are not confined to a wheelchair, an exoskeleton can provide the mobility and independence that they need to maintain a normal lifestyle. Exoskeletons have appeared in a number of science fiction movies, most notably assisting Sigourney Weaver in the 1979 "Alien" movie. However, these devices are available in much less threatening and less industrial forms to assist people with healthcare needs.

The ReWalk exoskeleton is a battery-powered system composed of a light, wearable exoskeleton with motors at the hip and knee joints. It controls movement using subtle changes in the patient's center of gravity. A forward tilt of the upper body is sensed by the system, which initiates the first step. Repeated body shifting generates a sequence of steps that mimic the functional natural gait of the legs.

Ekso Bionics has applied its clinical and engineering expertise to develop exoskeleton robotics for rehabilitation centers. The device itself is an exoskeleton, but it is used in rehabilitation as described in a later section. Patients post-stroke, brain injury, or spinal cord injury and those affected by MS can use exoskeletons in therapy to regain basic movements or even the ability to walk again. The patient may experience an increase in range of motion and the activation of muscles they had difficulty with before. Physical therapists remain a part of this treatment, leveraging the exoskeleton to improve their patients' gait and get them back to social and work functions.

Prostheses

More than one million annual limb amputations are carried out globally due to accidents, war casualties, cardiovascular disease, tumors, and congenital anomalies. Robotic prosthetic limbs integrate advanced mechatronics, intelligent sensing, and control for achieving higher-order lost sensor-motor functions while maintaining the physical appearance of the amputated limb. Robotic prosthetic limbs are expected to replace the missing limbs of an amputee restoring the lost functions and providing a natural esthetic appearance. These robotic prostheses contribute to enhanced social interaction, improved independent living, and productive work in society. Advances in electro-neural connectivity and miniaturization of mechanical motors will continue to make these limbs more effective and less expensive.

A Cleveland Clinic-led research team has engineered a first-of-kind neurorobotic prosthetic arm for patients with upper-limb amputations that allows wearers to think, behave, and function more like a person without an amputation. The robotic prosthetic combines intuitive motor control, a sense of touch, and grip kinesthesia (the intuitive feeling of opening and closing the hand).

Dr. Paul Marasco, Associate Professor in Cleveland Clinic Lerner Research Institute's Department of Biomedical Engineering, said, "We modified a standard-of-care prosthetic with this complex bionic system that enables wearers to move their prosthetic arm more intuitively and feel sensations of touch and movement at the same time. These findings are an important step toward providing people with amputation a complete restoration of natural arm function."

Companions

The PARO Therapeutic Robot looks like a baby harbor seal and is designed to provide the benefits of animal therapy without relying on live animals. Animal therapy is a common tool for easing patient stress, but there are not always trained animals available to satisfy current needs. The PARO robot is frequently used with elderly patients with dementia and has been proven to reduce stress and provide comfort to anxious patients. The soft, fuzzy device can respond to its name, enjoys being stroked, and, over time, develops a customized, pleasing personality tailored by its memory of previous interactions. PARO also naps, blinks, wiggles its flippers, and makes funny noises for its owner.

Devices like Connected Living's Temi robot also offer a window into the condition and safety of the elderly or patients needing home care. When equipped with visual, aural, and vocal sensors, these devices allow monitoring for alerts, consultations with clinicians, and reduce some of the fears of being entirely alone in a home. These companions combine the services of therapy animals, first alert buttons, and smart speakers to create a safer environment for independent living.

Mobile and Remote Services

Hospitals are very large and complex facilities in which mobility and a working knowledge of the locations of departments and services are required. Robots with mobility, intelligence, and independence can make significant contributions to the efficiency of these facilities.

Sterilization

Along with minimizing medical and surgical errors, hospital-acquired infections (HAIs) are a widespread problem in healthcare that could be improved with robots. The CDC reported that there were 722,000 HAIs in US acute care hospitals in 2011. These often occur because hospitals can't always clean rooms to 100% sterility between patients, whether due to time constraints or the simple invisibility of germs. Patients who are already immunocompromised are more susceptible to bacterial infection. Several companies have created room sterilization robots that are being used to reduce HAIs.

The Xenex LightStrike robot was one of the first of these. It uses full-spectrum UV rays to kill a range of infectious bacteria in an entire room. Significantly, it is effective against Coronavirus and Methicillin-resistant *Staphylococcus aureus* (MRSA). The robot generates bursts of high-intensity, short-wavelength ultraviolet (UV) light to kill disease-causing pathogens of all types. Light from the sun includes UVA and UVB rays, which can make it through the Earth's ozone layer, as well as UVC rays, which cannot. Because viruses and bacteria were not previously exposed to UVC, they never developed defenses against it. As a result, UVC light deactivates these germs and prevents them from reproducing. One five-minute light treatment from the robot is enough to destroy all viruses and bacteria within a two-meter area. To protect the safety of staff, the robot operates behind closed doors, with no people in the room. After the UVC light turns off, it is safe for housekeeping staff to enter the room and clean it without the risk of exposure to coronavirus or any other infectious germs. UVC robots are also produced by Clorox, Philips, Adibot, and BlueBionics.

Logistics

Hospitals are large and complex facilities. They require the delivery of supplies and medications throughout the day. This can be challenging for human staff who must learn the layout of the entire facility and spend the entire day looping through the hallways. A robot programmed with a map of the facility and equipped with appropriate sensors and decision logic is an attractive alternative to human delivery services.

MedStar Washington and other hospitals have employed the Aethon TUG robot to deliver food, supplies, medicines, and medical specimens. The device knows the layout of the hospital, is equipped with sensors to allow it to avoid people and obstacles, and has an electronic connection that allows it to call the elevator for a ride. MedStar Washington's six TUGs traveled 1,554 miles and made 13,800 stops, delivering medications, linens, and other essentials in a single year.

Navigation Assistance

As described in previous sections, navigating a large, complex hospital is a challenge for the staff that work there, and even more so for the thousands of patients, family members, and visitors that flow through a facility every day. Hospitals address this with lobby help desks, facility maps, and the assistance of all of their employees. Many hospital systems inculcate a culture among all staff of stopping to help lost visitors. Recognizing the behaviors of these lost visitors and offering assistance is an expected service from everyone who works in the facility.

Navigation robots can make a contribution to this service. Using voice recognition, voice generation, onscreen displays, and a perfect map of the facility, these devices can be located anywhere in the hospital and roam to where visitor flow is the heaviest. They can show and explain a route to the visitor's destination, or they can connect to a human assistant who can speak through the robot for more complicated problems. When showing the route onscreen is not sufficient, the robot can physically escort the visitor all the way to their destination.

Telemedicine

Telemedicine robots enable clinicians to see, hear, and speak with patients as if they were at their bedside, even when they are miles away at a different hospital or working from a home office. These devices provide a clear view of the patient, which is almost as good as being there. The cameras on the devices can be customized for the services to be provided. These may be high resolution with the ability to read monitors and charts in the room, possess magnification for examining small features like eye dilation and moles, or use infrared for viewing a patient's thermal profile.

During the COVID-19 pandemic, Liverpool Women's Hospital and Alder Hey Children's Hospital in the UK used two telemedicine robots to keep a full neonatal service running with limited staff and limited access. Other hospitals have used them to augment patient rounds by including physicians and specialists who would not otherwise be available during evening shifts.

Dr. Steve Jackson, chief medical information officer at University Hospitals of Leicester, has described the advantages that allow him to offer services at multiple locations during the same day, without the need to commute between them. Alternatively, physicians often dedicate whole days to a single facility and may not be able to serve patients at a facility until a later day. Jackson says, "I can consult with patients about whom my ward team are concerned or who might arrive on the ward between my ward round times and who might be able to go home outside of my designated ward round times. Such ad-hoc ward 'visits' would not otherwise be possible owing to my busy schedule and cross-site working."

He also believes that "From the point of view of having patients who need super specialist opinions, which I've used the tech for, it's absolutely superb. It means that patients don't have to wait to be transferred across the city to go to a different hospital because the specialists can consult with them there and then and make a plan which everyone's happy with."

Devices like Teladoc's InTouch, Vecna Technologies' Vgo, and the Double Robot are examples of telemedicine robots in use today.

Hazardous Contact

COVID, SARS, Ebola, and similar outbreaks created situations in which it was actually life threatening for human clinicians to be in contact with patients. When new diseases break and the transmission pathways, effects, and lethality of the disease are not yet well-known, it is a significant challenge to apply the necessary PPE, isolation, and contact procedures to keep clinicians safe while also meeting the needs of the patients. Some of the necessary interactions can be carried out through a robotic intermediary. Mobile, computerized devices can take patient history and some basic vital signs, thereby eliminating several of the pathways for disease transmission, making treatment centers safer for clinicians without withholding essential services from sick patients.

The InTouch, described earlier, was initially developed for patient-focused telemedicine applications but was adapted for hazardous contact specifically to address Ebola treatment dangers.

Training

Clinician educators have relied on manikins and part-task trainers for decades. These very useful devices reduce the reliance on live human actors, cadaveric tissue, and animals as surrogates for patients. In recent years these devices have become much more active and intelligent at providing these services.

Smart Manikins

Simulation manikins first emerged as life-sized cloth dolls for nurse training in 1911. Since then, they have been through multiple evolutions such that the most advanced of these are now also classified as robots for training. These allow clinicians to practice their healthcare skills on the robot before attempting to assess or perform a treatment on a human patient. Using smart robotic manikins, learners can be better prepared to manage their immediate reactions to critical situations. These devices also reduce the need for supervision or direct oversight during many clinical patient encounters. Robotic manikins allow clinicians to rehearse repeatedly under identical conditions and support the collection of objective data that measures improved performance.

Laerdal, founded in 1940 and headquartered in Stavanger, Norway, produces a number of manikin products that vary in age, gender, and clinical treatment applications. Devices that would be considered robots include SimMan, SimMom, SimJunior, SimNewB, and Premature Anne.

These devices can display neurological and physiological symptoms that clinicians must diagnose and treat. Laerdal's robot manikins have a pulse, blood pressure, breath, lung sounds, heart sounds, pulse oximetry, and a monitor that can display EKG, arterial waveforms, and pulmonary artery waveforms. These manikins can be used to train tasks like CPR, bag-mask ventilation, intubation, defibrillation, chest tube placement, and others. Similar devices are available from Gaumard, CAE Health, and others.

Anatomical Replicas

Just as simple full-body manikins evolved into robotic devices, partial-body replicas are following the same path. Most part-task trainers are simple latex-covered frames that present a passive piece of anatomy for learning and rehearsing a specific procedure. But a few devices like the Gaumard Advanced CPR and Airway Trainer are adding smart dynamic actions to these replicas. As computers and sensors become smaller and cheaper, this opens the door for incorporating them into lower-cost and more focused training devices. Like the full-body manikins, the objective is to replicate the functions of the human body and the responses that occur during a medical procedure. The device is no longer an inanimate piece of latex, but a robotic replica of the anatomy and its dynamic functionality.

Industrial Automation

All manufacturing and warehouse industries have been using robots for decades. These devices perform well defined and repetitive tasks much more quickly, accurately, and economically than their human counterparts. In the healthcare space, similar manifestations of these robots are found in pharmacies and laboratories.

Pharmacy

Robotic picking arms can be used in both large pharmacy warehouses and in smaller dispensaries at individual hospitals. The principles of operation are very similar to the plethora of warehouse robots that are used for every other type of order fulfillment. The devices are programmed with the location of the drug that has been requested and then read barcodes when they arrive at the destination to ensure that they are retrieving the right product.

These machines tend to employ either chaotic or channel-fed dispensing systems. Omnicell's Cyrus Hodivala says,

"In chaotic storage, the robot takes a pack and puts it on the shelves, and only the robot knows where that pack has gone. It does this for space efficiency, tessellating packs together on a shelf to fit the most packs inside that it possibly can. That's great if you've got a very big formulary, which is why hospitals love it, because you've got thousands of drugs on register."

The main problem with chaotic storage is that the picking speed can be quite slow because the robot will often need to move packs around to find the one it needs. The time between making a dispensing request to receiving the pack in a chaotic storage system is around 14 seconds.

The alternative method of channel-fed dispensing involves storing drugs in individual metal or plastic channels, each of which is dedicated to just one drug type. Channel-fed dispensing allows for much faster picking speeds compared to chaotic storage, taking around four seconds to dispense a single pack.

Hodivala says: "The downside for channel-fed is your channels need to be sized for your packs. If you've got lots of varying pack sizes, you must continually keep reallocating channels. Also, if you're only holding one pack of a particular drug type an entire channel needs to be dedicated to it, so it's not very space efficient."

The University of California, San Francisco (UCSF) hospital, also uses robots to process daily medication orders for individual patients. Using mechanical arms, they select and sort multiple medications into barcoded packets that are customized to the patient.

Laboratory

Clinical laboratories use small, nimble robots to perform the most repetitive laboratory tasks, relieving humans from this inhuman work. ABB Robotics, which runs a research center at the Texas Medical Center (TMC) Innovation Institute in Houston, estimates that the market for laboratory robots in healthcare will reach 5,000 by 2025.

This influx of robot assistants doesn't mean humans will be banished from the laboratory, rather the robots will be supporting humans with dangerous or dull activities. Currently, robots are used to organize the most routine and repetitive part of the laboratory, like centrifuging, aliquoting, and automating routine chemistry, immunoassay, hematology, and urinalysis. The systems are guided by barcodes that tell the robots in various instruments what to do. This is most common for routine high-volume tasks that can justify the purchase of a robotic device. But, once the robots are on premises, there is the potential to add end effectors and software to allow them to perform lower volume, more specialized tasks as well. As these robots become more sophisticated, more accurate, more intelligent, and cheaper, they become a more attractive alternative to human laboratory technicians.

Advantages

Robotic devices in healthcare have evolved from early experiments to fully integrated contributors to healthcare services and hospital operations. Each of the categories and examples shared in this chapter carry with it at least one advantage in the clinical practice and business operations of a healthcare system (Figure 6.5). The most prominent of these include the following:

- Improving the affordability of services
- Extending the reach of clinicians

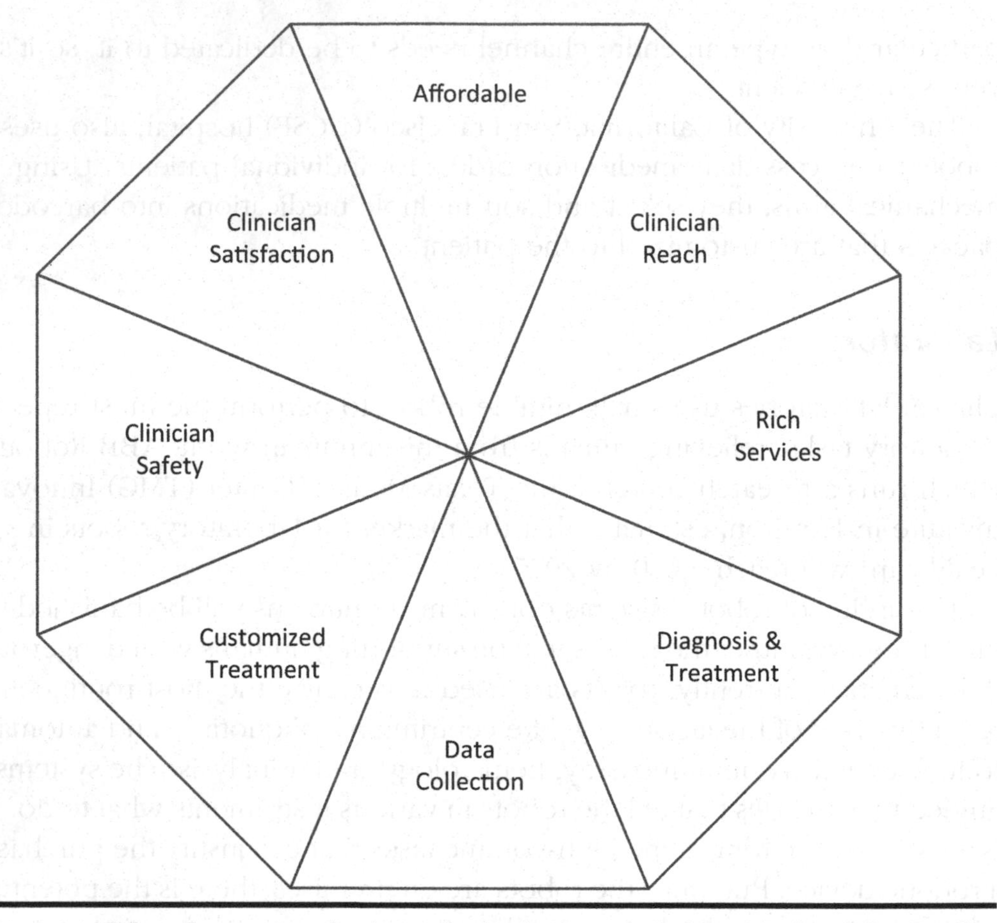

Figure 6.5 Eight advantages of using robotic devices to assist with the delivery of healthcare services.

- Increasing the richness of services to patients
- Improving the diagnosis and treatment of conditions
- Improving data collection and metrics
- Customizing treatment for each patient
- Increasing safety for clinicians
- Increasing clinician satisfaction

Disadvantages and Concerns

Along with the advantages of adding robotics to the service ecosystem, there are a few disadvantages or concerns that are coupled with them. Those most often cited are as follows:

- Less human connection to the patient
- High capital investment cost
- Creating a dependence on robots for essential functions

Adoption Factors

Business leaders and clinicians often cite the factors that they believe will be important as robotic devices continue to be offered for use in healthcare (Figure 6.6). These leading requirements for adoption are as follows:

- Clinical effectiveness
- Capability and functionality
- Safety and reliability
- Usability
- Cost effectiveness

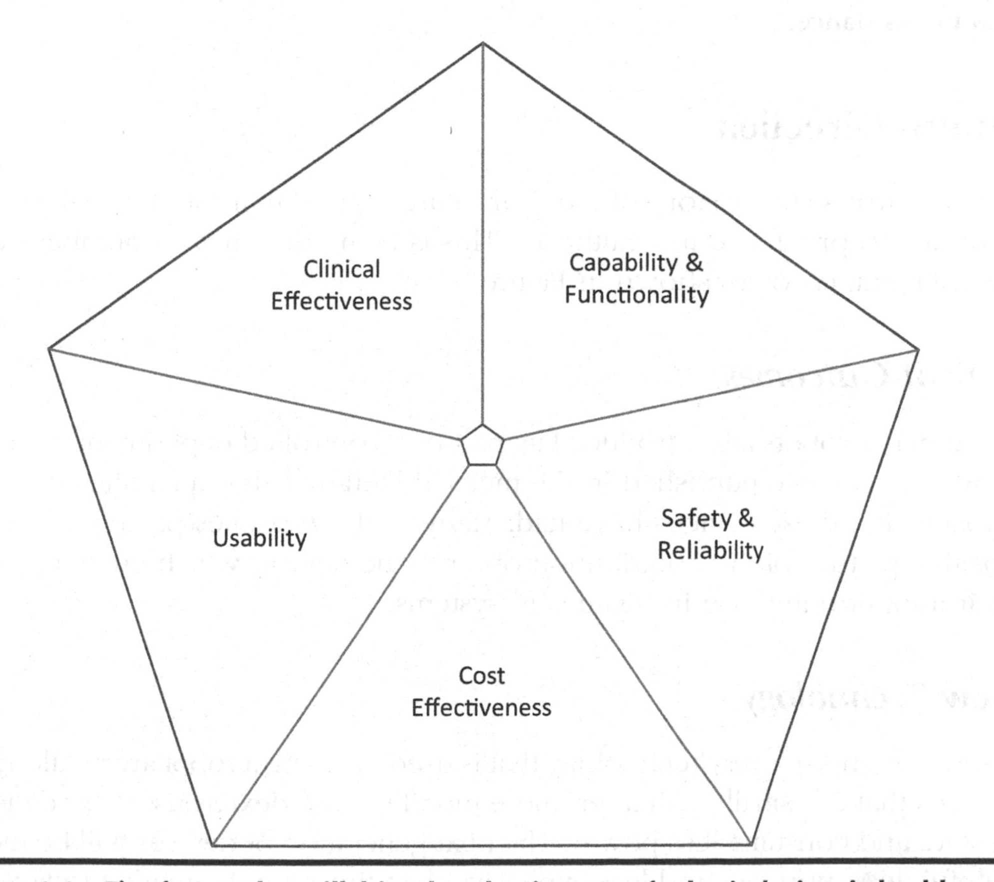

Figure 6.6 Five factors that will drive the adoption rate of robotic devices in healthcare.

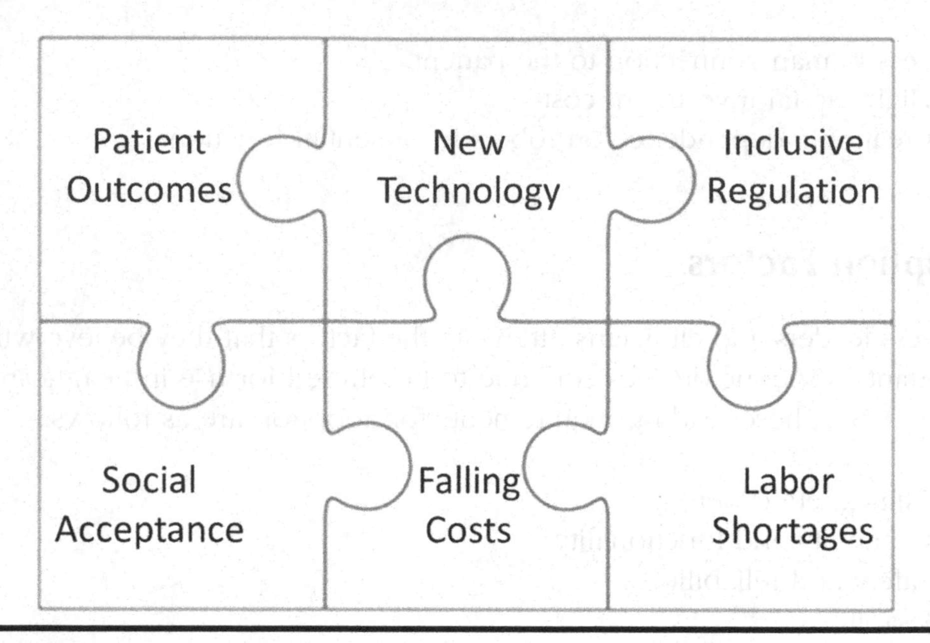

Figure 6.7 Six future forces that create a healthcare environment that is inclusive to robotic assistance.

Future Direction

It is no surprise that visions of the near future suggest that robotic devices will continue to proliferate in healthcare. This is being driven by a number of complimentary forces shown in Figure 6.7.

Patient Outcomes

Since many robots are introduced as part of a controlled experiment, there is growing evidence published in the medical literature that quantifies the impacts that these devices have had. Many of these studies present a very positive picture of the contribution of a robotic device, which encourages their adoption and use in healthcare systems.

New Technology

Improvements in every technology that is used to create a robot are resulting in devices that are smaller, smarter, more mobile, more dexterous, more connected, and consume less power. The next generation of devices will be more helpful, less intrusive, and less expensive than those on the market today.

Inclusive Regulations

The healthcare field has been limited by regulations on the use of telemedicine and assistive devices for decades. One impact of the COVID crisis has been the necessity to connect patients with healthcare providers in new ways. This has opened the door on regulation and insurance reimbursement such that robots and similar technologies can be billed to these payors, reducing the burden on patients and end-users.

Social Acceptance

The term "robot" has historically conjured threatening images of the Terminator. But, with exposure to the many forms of robots and media stories about their benefits, the general public is becoming less fearful and more accepting of the devices. Contributing to this acceptance is the maturation of a generation of children who grew up with these technologies, are comfortable using them, and have aged into positions of leadership in healthcare, government, and medical device companies.

Labor Shortages

The demand of healthcare services is exceeding the ability of an all-human workforce to deliver these in the traditional manner. Meeting the needs of an aging population requires the recruitment of a new set of hands to assist the limited supply of healthcare providers. Some of these hands can come from smart robotic devices.

Falling Costs

Improved technologies and loosened regulations create an environment in which robotic devices can be produced in larger numbers, reducing the per unit price. Robotic assistance in healthcare will become less costly as it becomes more prevalent and better integrated into the larger healthcare delivery system.

The advantages of adding robotic devices to the healthcare system are overcoming the resistance to the devices that were a natural part of a new technology in its early stages and which were poorly understood by both clinicians and patients. These devices are no longer relegated to science fiction stories but have become a common component of a modern, integrated healthcare system that delivers the best outcomes for patients.

Bibliography

AdventHealth (2022). Robotics in healthcare: past, present, and future. https://online.ahu.edu/blog/infographic/robotics-in-healthcare/

Alvey, R (2021). Robotics in healthcare. *Online Journal of Nursing Informatics (OJNI)*, 25(2). https://www.himss.org/resources/online-journal-nursing-informatics

Balasubramanian, S, Klein, J, Burdet, E (December 2010). Robot-assisted rehabilitation of hand function. *Current Opinion in Neurology*, 23(6):661–670 doi: 10.1097/WCO.0b013e32833e99a4

Banks, M (2022). How robots are redefining healthcare: 6 recent innovations. *Robotics Tomorrow*. https://www.roboticstomorrow.com/story/2022/03/how-robots-are-redefining-health-care-6-recent-innovations/18339/

Bieller, D (2021). IFR. The role of robots in healthcare. https://ifr.org/post/the-role-of-robots-in-healthcare-part2

Brainlab (2021). The types of medical robots in use today and in the future. https://www.brainlab.com/journal/types-of-medical-robots-in-use-today-and-in-the-future/

Case Western Reserve. 5 Medical robots making a difference in healthcare. https://online-engineering.case.edu/blog/medical-robots-making-a-difference

DelveInsights (2021). How robots are introducing a new dimension to healthcare service delivery. https://www.delveinsight.com/blog/robotics-in-healthcare

Dolic, Z et al. (2019). Robots in healthcare: a solution or a problem? *European Parliament Workshop on AI in Healthcare*. https://www.europarl.europa.eu/RegData/etudes/IDAN/2019/638391/IPOL_IDA(2019)638391_EN.pdf

Giansanti, D (2020 Dec 30). The rehabilitation and the robotics: are they going together well? *Healthcare*, 9(1):26. doi: 10.3390/healthcare9010026. PMID: 33396636; PMCID: PMC7823256.

GlobalData (2020). What are the main types of robots used in healthcare?. https://www.medicaldevice-network.com/comment/what-are-the-main-types-of-robots-used-in-healthcare/

Gorgey, AS (2018 Sep 18). Robotic exoskeletons: the current pros and cons. *World Journal of Orthopedics*, 9(9):112–119. doi: 10.5312/wjo.v9.i9.112. PMID: 30254967; PMCID: PMC6153133.

Hockstein, NG, Gourin, CG, Faust, RA, Terris, DJ (2007). A history of robots: from science fiction to surgical robotics. *Journal of Robotic Surgery*, 1(2):113–118. doi: 10.1007/s11701-007-0021-2. Epub 2007 Mar 17. PMID: 25484946; PMCID: PMC4247417.

Intel. Robotics in healthcare: The future of robots in Medicine. https://www.intel.com/content/www/us/en/healthcare-it/robotics-in-healthcare.html

Jibb, LA, Birnie, KA, Nathan, PC, Beran, TN, Hum, V, Victor, JC, Stinson, JN (2018). Using the MEDiPORT humanoid robot to reduce procedural pain and distress in children with cancer: a pilot randomized controlled trial. *Pediatric Blood and Cancer*, 65(9).

Joseph et al. (2018). A review of humanoid robotics in healthcare. MATEC Web Conference.

Kyrarini, M, Lygerakis, F, Rajavenkatanarayanan, A, Sevastopoulos, C, Nambiappan, HR, Chaitanya, KK, Babu, AR, Mathew, J, Makedon, F (2021). A survey of robots in healthcare. *Technologies*, 9(1):8. 10.3390/technologies9010008

Marín-Méndez, H, Marín-Novoa, P, Jiménez-Marín, S, Isidoro-Garijo, I, Ramos-Martínez, M, Bobadilla, M, Mirpuri, E, Martínez, A (2021). Using a robot to treat non-specific low back pain: results from a two-arm, single-blinded, randomized controlled trial. *Frontiers in Neurorobotics*, 15, DOI=10.3389/fnbot.2021.715632, https://www.frontiersin.org/article/10.3389/fnbot.2021.715632

Minjun, K, Anak, AJ, Cheang, UK (2017) *Microbiorobotics* (Second Edition). Elsevier,

Mohammad, S (2013 Jan-Apr). Robotic surgery. *Journal of Oral Biology and Craniofacial Research*, 3(1):2. doi: 10.1016/j.jobcr.2013.03.002. PMID: 25737871; PMCID: PMC3941295.

Ornes, S (2017). Medical microrobots have potential in surgery, therapy, imaging, and diagnostics. *Proceedings of the National Academy of Sciences of USA*, 114(47):12356–12358 10.1073/pnas.1716034114

PriceWaterhouseCooper (2017) What doctor? *Why AI and robotics will define New Health*. https://www.pwc.com/gx/en/industries/healthcare/publications/ai-robotics-new-health/transforming-healthcare.html

Riek, L (2017). Healthcare Robotics arxiv.org. https://arxiv.org/pdf/1704.03931.pdf

Taylor, RH, Simaan, T, Menciassi, A, Yang, G (July 2022). Surgical robotics and computer-integrated interventional medicine [Scanning the Issue]. *Proceedings of the IEEE*, 110(7), 823–834, doi: 10.1109/JPROC.2022.3177693.

Tietze & McBride, Robotics and the Impact on Nursing Practice (2019). American Nurses Association. https://www.nursingworld.org/~494055/globalassets/innovation/robotics-and-the-impact-on-nursing-practice_print_12-2-2020-pdf-1.pdf

Zhou X, et al. Artificial intelligence in surgery. https://arxiv.org/pdf/2001.00627.pdf

Chapter 7

RFID in Healthcare

Shaan Revuru

Radio-frequency identification (RFID) technology has been around for decades and is still going strong today. It has been gaining attention in healthcare lately due to its contribution to help improve patient safety, increase operational efficiency, and enhance the effectiveness of clinical processes. This chapter will provide a comprehensive overview of RFID technology, from its inception during World War II to present-day applications in the healthcare industry. We will discuss the different types of RFID tags, readers, and antennas, as well as how RFID is used for tracking, tracing, and monitoring assets, inventory, and people and preventing loss in hospitals. In this chapter, we will give a comprehensive overview of RFID technology and how it is being used in the healthcare industry.

The History of Radio-Frequency Identification Technology

Radio-frequency (RF) technology, one of the most important wireless technologies available today, has advanced significantly since its beginnings in the early twentieth century. While the Russian physicist Leon Theremin (n Lev Sergeyevich Termen) is commonly attributed with the creation of the first RF device and the first successful application of RF technology in 1946 [1], radio-frequency identification (RFID) as a technology has much earlier roots.

At the center of RFID technology are two key technologies, the radar technology and radio broadcasting technology. The first basic radar system was initially developed at the US Naval Aircraft Radio Laboratory in 1922 by

DOI: 10.4324/9781032690315-7

Albert H. Taylor and Leo C. Young, however, it wasn't until 1930 that Lawrence A. Hyland with Albert H. Taylor and Leo C. Young at the U.S. Naval Research Laboratory developed a mechanism using radio signals to detect passing aircraft and obtained a patent [2]. In 1935, a British patent was issued to Sir Robert Watson-Watt for a Radar System for Air Defense [3]. The foundation of any radio broadcasting technology, electricity, and magnetism, however, was established in the early nineteenth century [4]. Throughout the 1940s, several variations of this technology were explored, the most common transponder system being "Identification, Friend or Foe: IFF" for identifying aircrafts during World War II [5,6].

The publication of RFID as a concept is attributed to Harry Stockman, who in 1948 wrote the paper "Communications by means of reflected power" [7]. In this paper, Harry Stockman recognized there was still a lot of work needed for the RFID technology to become mainstream and for it to be applied in everyday life.

After the war, RFID technology remained primarily in military and government applications, in 1973, Charles A. Walton received a patent for a keyless door entry system using passive RFID [8]. Later in the 1970s, Los Alamos National Laboratories began experimenting using RFID for commercial purposes, and in 1983, Los Alamos filed for a patent on an "Active Badge System," which was a badge that emitted a signal that could be tracked by a reader system. The badges were initially used to track the location of employees in buildings so that they could be easily located in case of an emergency. In 1987, Philips Electronics introduced the first commercially available RFID tag, called the "Electronic Product Code" (EPC). The EPC was designed to be used in inventory management and tracking applications.

Though RFID tags were used in the 1970s to monitor railway cars, it was not until the late 1990s that RFID became widely used in commercial applications. In 1999, the first large-scale deployment of RFID tags was conducted by the U.S. Air Force, which placed RFID tags on all aircraft at their bases worldwide. This was followed by similar deployments by other military organizations, such as the United Kingdom's Ministry of Defense.

One of the reasons RFID technologies did not take off in commercial applications until the late 1990s was the cost of RFID tags. The early EPC tags cost around $0.50 each, which made them too expensive for most commercial applications. However, tag prices began to drop in the early 2000s, and by 2003, EPC Gen-II tags were available for less than $0.20 each. This decrease in price led to an increase in the adoption of RFID technology in the retail sector, with companies such as Walmart and Target deploying RFID tagging of products in their stores.

How Does RFID Work?

RFID technology has been around for over 60 years, but it was only recently that it began to be used extensively in a variety of applications. One of the most important factors driving the growth of RFID is its ability to track moving objects in real time. This is particularly valuable in industries such as healthcare, where tracking assets is essential to providing quality care.

RFID technology is based on the use of electromagnetic fields to automatically identify and track tags attached to objects. The tags contain electronically stored information, which can be read from a distance by an RFID reader.

There are three main components of an RFID system:

Tags: Also called transponders, these are affixed to the objects that need to be identified and tracked. Tags come in various forms, including labels, cards, fobs, and coins.
Readers: RFID readers emit radio waves that activate the tags and retrieve the stored data. Readers can be handheld devices, mobile devices, desktop devices, or fixed installations.
Antennas: Antennas transmit the radio waves from the reader to the tag and vice versa.

RFID technology is also being used increasingly for patient safety and quality assurance purposes. For example, RFID tags can be affixed to medication bottles or IV pumps to ensure that the correct medication is being administered to the correct patient at the correct time. In addition, RFID-enabled medical records can help staff quickly locate a patient's chart in an emergency situation.

RFID Readers and Antennas

RFID readers and antennas are the two main components of an RFID system. RFID readers are used to read data from RFID tags, while RFID antennas are used to transmit and receive data between RFID tags and readers.

RFID Reader Types

There are three main types of RFID readers: mobile or handheld, fixed, and networked.

Figure 7.1 Mobile Handheld RFID reader.

Source: Cipherlab – RS35 UHF RFID Reader

Mobile or Handheld Readers are portable handheld devices that can be carried around by a user and used to track the location of assets in real-time and to read data from RFID tags. These readers are often used in retail applications, such as inventory management (Figure 7.1).

Fixed readers are stationary devices that are mounted on a surface, such as a wall or a ceiling, are usually installed at entry and exit points, such as doors or gates, and are used to track the movement of tags in and out of a particular area. These readers are often used in industrial applications, such as tracking assets in a warehouse (Figure 7.2).

Networked readers are connected to a network and can be accessed by multiple users. These readers are often used in enterprise applications, such as asset tracking (Figure 7.3).

Active, Passive, and Semi-Active RFID Readers

Depending on the type of tags used RFID Readers can be categorized into active, passive, or semi-active RFID Readers.

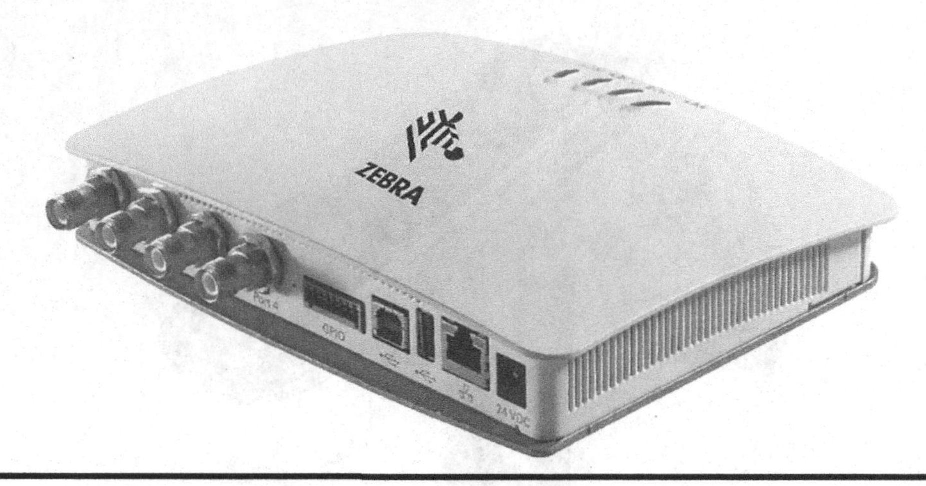

Figure 7.2 Fixed RFID reader.

Source: Zebra – FX7500 Fixed RFID Reader

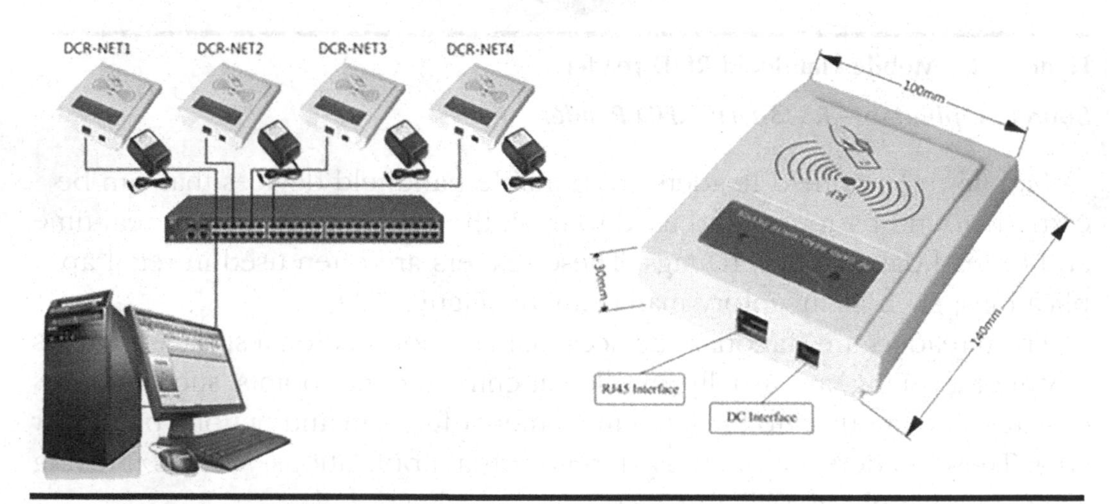

Figure 7.3 Networked RFID reader.

Source: Emartee – Network RFID Reader V3, 125 KHz HID Card

Active RFID Readers

Active RFID readers use battery-powered tags to communicate with the reader. The active RFID Reader transmits a signal to active RFID Tag, active RFID Tag transmits back to the active RFID Reader a response signal that contains the tag's unique identification number. Additionally, active RFID Readers use battery-powered Active RFID Tags to achieve readability from great distances. The active RFID Readers are typically expensive, but they

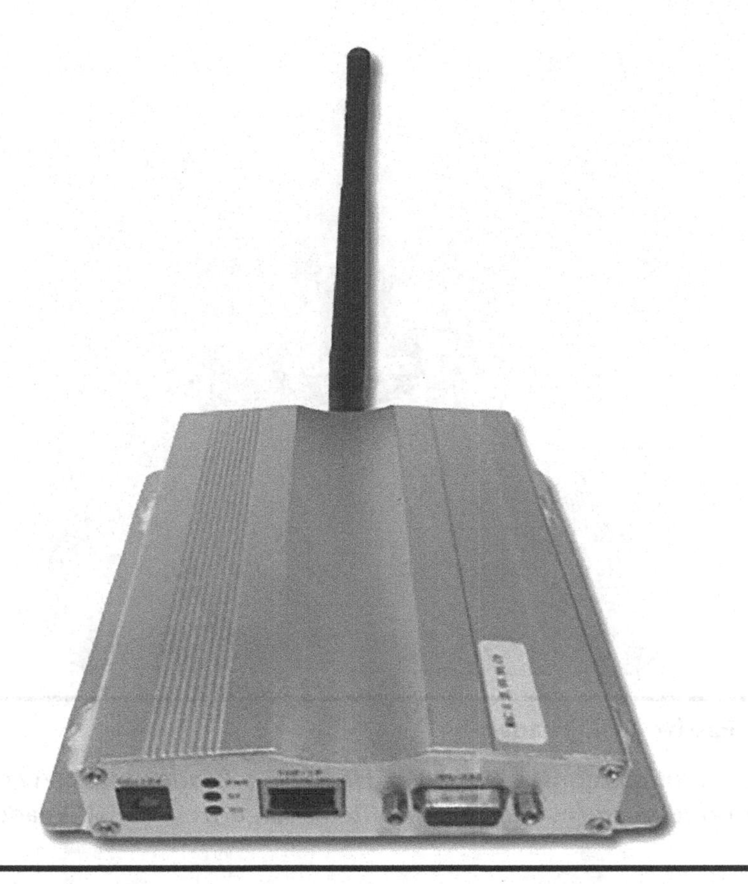

Figure 7.4 Active RFID reader.

Source: gaorfid.com – 2.45 GHz Active RFID Reader https://gaorfid.com/wp-content/uploads/2019/06/2.45-GHz-Active-RFID-Reader-Kit-for-Access-Control.jpg

offer great ranges and advanced functionality. Active RFID Readers are often used in industrial applications where long-range communication is required (Figure 7.4).

Passive RFID Readers

Passive RFID Readers rely on passive RFID Tags to communicate with the Passive RFID Readers. The Passive RFID Reader transmits a signal to the Passive RFID Tag, the Passive RFID Tag transmits back to the Passive RFID Reader a response signal with the tag's unique identification number. Since the Passive RFID Readers use Passive RFID Tags that are not battery powered and instead rely on the energy of the signal transmitted by the Passive RFID Reader, they are

Figure 7.5 Passive RFID reader.

Source: tresrfsolutions.com – TRES900 UHF Passive Reader https://www.controlledproducts.com/ecomm_images/items/large/tres-900%20reader.jpg

typically less expensive, however, this reduction in cost results in lower read ranges and lack of any advanced functionality. Passive RFID Readers are typically used in low-cost applications where simple tracking or identification is required, like retail applications (Figure 7.5).

Semi-Active RFID Readers

Semi-active RFID Readers rely on battery power to enhance the range and sensitivity of the semi-active RFID system. While the passive RFID systems rely on energy emitted by the passive RFID Reader to activate the passive RFID Tag, the semi-active RFID systems use a battery-powered tag that transmits a signal with the tag's unique identification number frequently that is read by the semi-active RFID Reader. The battery in the semi-active RFID Tag provides power to the tag's transmitter that allows the tag to transmit its signal over a longer range than a Passive RFID Tag. The signal transmitted by the tag is picked up by the reader and is used to identify and locate the tag.

Semi-active RFID Readers are most commonly used where a longer read range is required, in applications like inventory tracking in a warehouse or monitoring how goods are moving in a supply chain. Semi-active RFID systems are also used in applications to track vehicles and other items in transportation and logistics.

RFID Antenna Types

There are four main types of RFID antennas: linear polarized, circular polarized, patch, and omnidirectional.

Linear polarized antennas have a horizontal or vertical orientation and are often used in retail applications (Figure 7.6).

Circular polarized antennas have a circular orientation and are often used in industrial applications (Figure 7.7).

Figure 7.6 Linear polarized RFID antenna.

Source: RFMAX Antennas – RLPA-902-10-NF

Figure 7.7 Circular polarized RFID antenna.

Source: RFMAX Antennas – R9028-LPF-SSF

Patch antennas are flat and are often used in enterprise applications (Figure 7.8).

Omni-directional antennas emit RF signals in all directions and are often used in tracking people or animals (Figure 7.9).

RFID Technology Types

The technology that is used to communicate between the tags and readers determines what sort of RFID tag it is. The RF range and the way a tag communicates with the reader are some of the criteria that distinguish one type of RFID tag from another (low, high, or ultra-high).

Just as there are different types of RFID tags, there are also different types of RFID technology. The three most common are active, passive, and semi-active. Active and Passive RFID tags use inductive coupling to communicate with a reader, while semi-active tags use both inductive coupling and a power source (such as a battery) (Figures 7.10 and 7.11).

Figure 7.8 Patch RFID antenna.
Source: RF Superstore – RFS-200057

Figure 7.9 Omni-directional RFID antenna.
Source: Laird Antennas – CAF95956

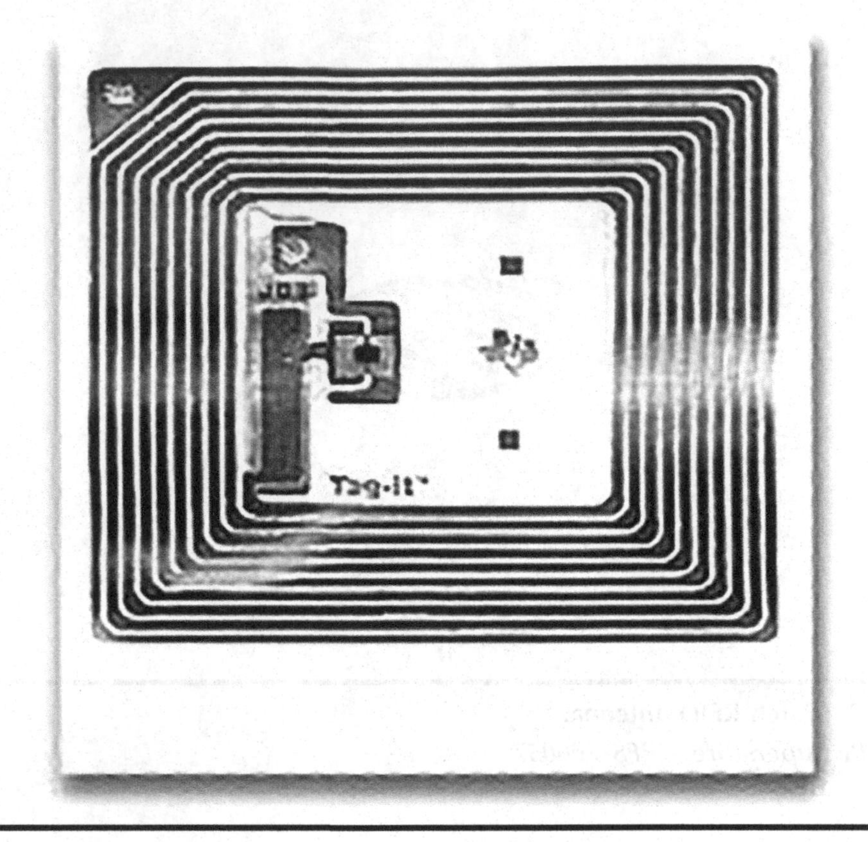

Figure 7.10 RFID chip surrounded by its long antenna coil.

Source: DLA

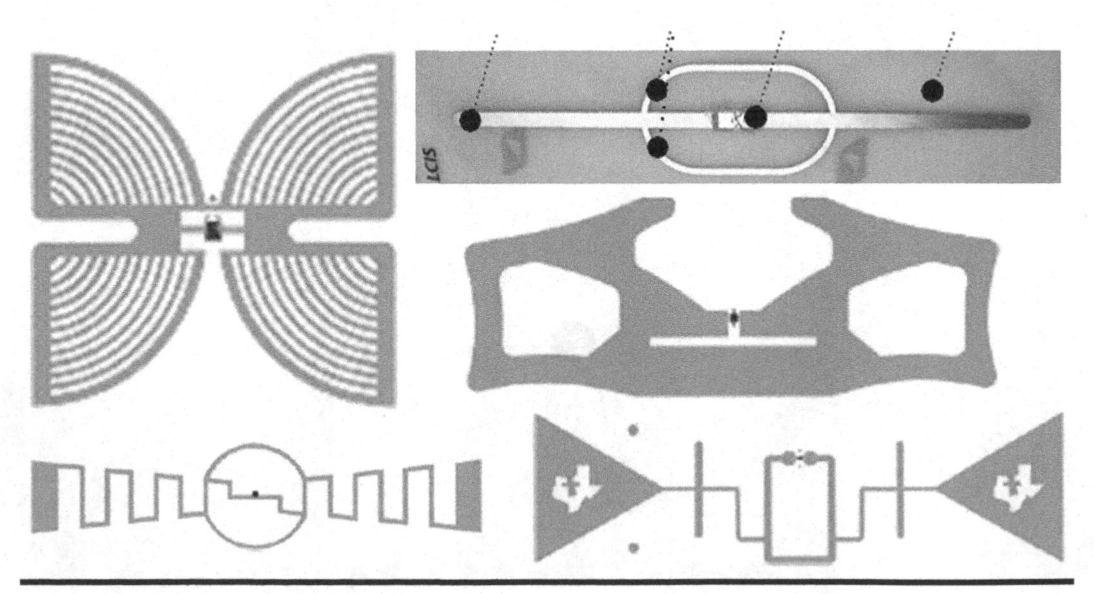

Figure 7.11 LF, HF, and UHF Tags.

Source: Web/Internet

Active RFID tags

Active RFID tags have their own power source, which allows them to transmit data over greater distances than passive tags. They also tend to be more expensive than other types of RFID tags. Semi-passive RFID tags rely on the power from the reader to transmit data, but they also have a power source that allows them to store data. This type of RFID tag is often used for applications that require more memory than what a passive tag can provide.

Advantages

- Long-read range: active RFID tags can be read at distances up to 100 m (328 ft.).
- No line of sight required: active RFID tags can be read even if they are not in the direct line of sight of the reader.
- Greater memory capacity: because they have their own power source, active RFID tags can store more data than passive tags.

Disadvantages

- High cost: active RFID tags are more expensive than other types of RFID tags.
- Short battery life: the batteries in active RFID tags only last for a few years.
- Maintenance required: because they have batteries, active RFID tags need to be regularly replaced or recharged.

Passive RFID Tags

Passive RFID tags rely on the power from the reader to transmit data. They do not have their own power source and are therefore much less expensive than active or semi-passive tags. Passive RFID tags are often used in applications where cost is a major concern, such as inventory management (Figure 7.12).

Advantages

- Low cost: passive RFID tags are much less expensive than active or semi-passive tags.
- No maintenance required: because they don't have batteries, passive RFID tags don't need to be regularly replaced or recharged.

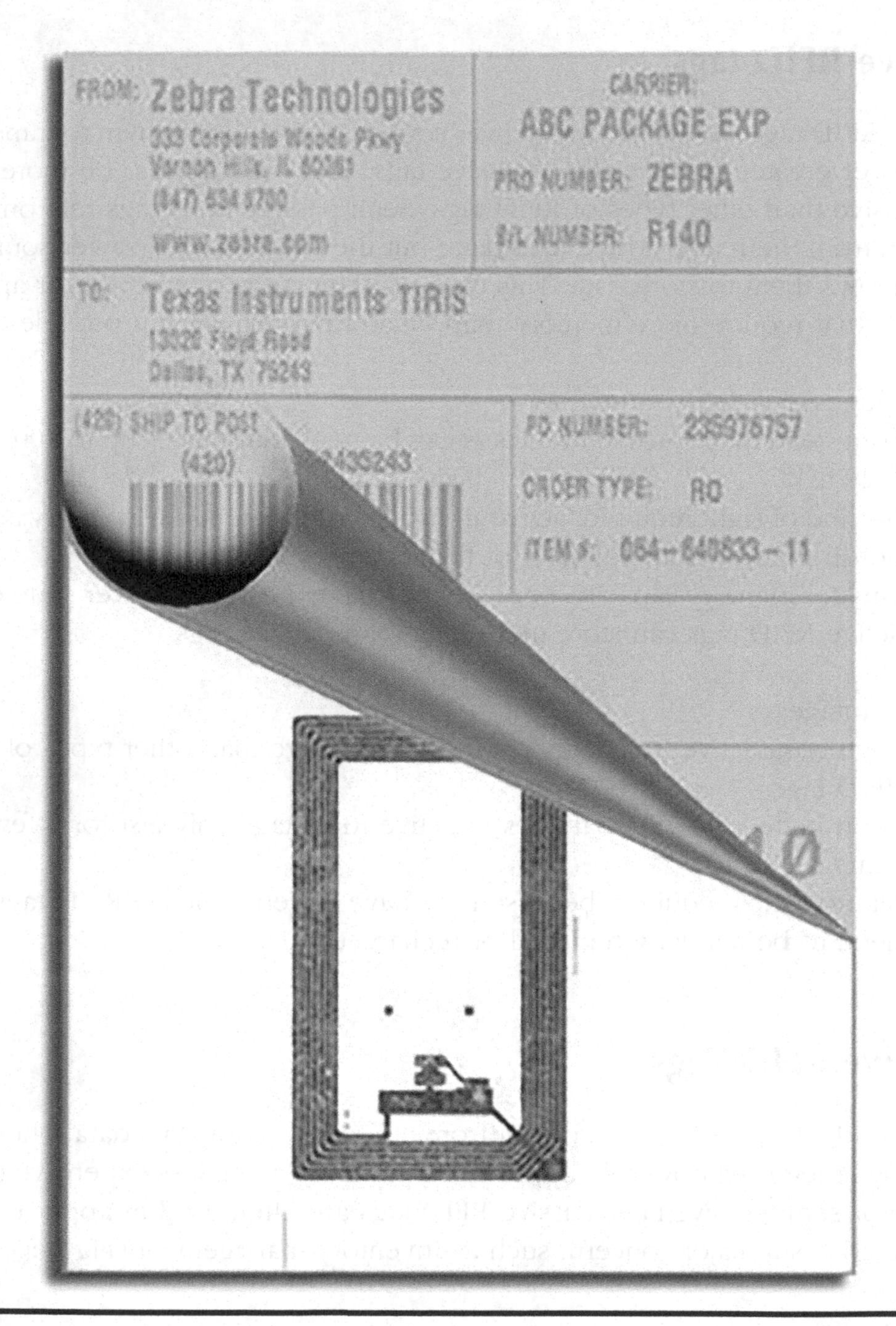

Figure 7.12 Passive RFID label.
Source: DLA

Disadvantages

- Short read range: passive RFID tags can only be read at distances up to a few meters.
- Line of sight required: passive RFID tags must be in the direct line of sight of the reader in order to be read.
- Limited memory capacity: because they don't have their own power source, passive RFID tags can only store a limited amount of data.

Semi-Active RFID Tags

As the name implies, semi-active RFID tags are a hybrid of active and passive RFID tags. They have their own power source, which allows them to transmit data over greater distances than passive tags, but they don't rely on the power from the reader to do so. Semi-active RFID tags are often used in applications where longer read ranges are required but the cost is still a concern.

Advantages

- Long-read range: semi-passive RFID tags can be read at distances up to 100 m (328 ft.).
- No line of sight required: semi-passive RFID tags can be read even if they are not in the direct line of sight of the reader.

Disadvantages

- Moderate cost: while not as expensive as active RFID tags, semi-passive RFID tags are more costly than passive RFID tags.
- Limited battery life: the batteries in semi-passive RFID tags only last for a few years.
- Maintenance required: because they have batteries, semi-passive RFID tags need to be regularly replaced or recharged.

RFID Tags According to Frequency

RFID tags can be classified according to the frequency range they use to communicate with a reader. Low-frequency (LF) RFID tags operate at 125 KHz and 134 KHz range, while high-frequency (HF) RFID tags operate at 13.56 MHz. Ultra-high-frequency (UHF) RFID tags, which are the most

Table 7.1 RFID Frequency Bands

Band	*Low Frequency*	*High Frequency*	*Ultra-High Frequency 900 MHz*	*Microwave*
Frequency	125 KHz 134.2 KHz	13.56 MHz	400 MHz 865–869 MHz Europe 902–928 MHz, USA 952–954 MHz, Japan	2.45 GHz 5.8 GHz
Applications	Security and Access Control, Keyless Badge Entry	Transport Applications, Electronic Tolling, Border Control – Electronic Passports	Supply Chain Logistics, Container Management, Inventory Management	Retail

common type of RFID tags, operate at frequencies ranging from 850 MHz to 950 MHz in North America and Europe, and from 920 MHz to 950 MHz in Asia. This frequency range is summarized in Table 7.1.

Low-Frequency (LF) RFID Tags

LF RFID tags are often used in applications where a low cost is more important than a long-read range. LF RFID tags typically have a read range of up to three feet (one meter). LF RFID tags are often used in access control, animal identification, and automotive applications.

Advantages

- Low cost: LF RFID tags are less expensive than HF and UHF RFID tags.
- No line of sight required: LF RFID tags can be read even if they are not in the direct line of sight of the reader.

Disadvantages

- Short read range: LF RFID tags have a relatively short read range, typically up to three feet (one meter).

High-Frequency (HF) RFID Tags

HF RFID tags are often used in applications where a longer read range is required. HF RFID tags typically have a read range of up to 30 feet (9 meters). HF RFID tags are often used in contactless payments, identity cards, and library books.

Advantages

- Long-read range: HF RFID tags have a longer read range than LF RFID tags, typically up to 30 feet (9 meters).
- No line of sight required: HF RFID tags can be read even if they are not in the direct line of sight of the reader.

Disadvantages

- High cost: HF RFID tags are more expensive than LF RFID tags.

Ultra-High-Frequency (UHF) RFID Tags

UHF RFID tags are the most common type of RFID tag. They are often used in applications where a long-read range is required. UHF RFID tags typically have a read range of up to 100 feet (30 meters). UHF RFID tags are often used in asset tracking, inventory management, and supply chain management.

Advantages

- Long-read range: UHF RFID tags have a longer read range than LF and HF RFID tags, typically up to 100 feet (30 meters).
- No line of sight required: UHF RFID tags can be read even if they are not in the direct line of sight of the reader.

Disadvantages

- High cost: UHF RFID tags are more expensive than LF and HF RFID tags.
- Maintenance required: UHF RFID tags need to be regularly replaced or recharged.

RFID Tags According to Memory Capacity

RFID tags can also be classified according to their memory capacity. There are three main types of RFID tags: read-only, write-once/read-many (WORM), and read/write.

Read-Only RFID Tags

Read-only RFID tags can only be written once and cannot be rewritten. Read-only RFID tags are often used in applications where the data on the tag needs to be permanent, such as in product identification.

Advantages

- Permanent: the data on read-only RFID tags cannot be rewritten, making it permanent.
- Low cost: read-only RFID tags are less expensive than WORM and read/write RFID tags.

Disadvantages

- Limited data storage: read-only RFID tags can only store a limited amount of data.
- Limited flexibility: the data on read-only RFID tags cannot be changed, so they are not well suited for applications where the data needs to be updated frequently.

Write-Once/Read-Many (WORM) RFID Tags

WORM RFID tags can be written to once and then read an unlimited number of times. WORM RFID tags are often used in applications where the data on the tag needs to be permanent but also needs to be accessible, such as in library books.

Advantages

- Permanent: the data on WORM RFID tags cannot be rewritten, making it permanent.
- High data storage: WORM RFID tags can store a large amount of data.
- Good flexibility: the data on WORM RFID tags can be read an unlimited number of times, so they are well suited for applications where the data needs to be accessed frequently.

Disadvantages

- High cost: WORM RFID tags are more expensive than read-only RFID tags.

- Limited data rewriting: the data on WORM RFID tags can only be written to once, so they are not well suited for applications where the data needs to be updated frequently.

Read/Write RFID Tags

Read/write RFID tags can be written to and rewritten multiple times. Read/write RFID tags are often used in applications where the data on the tag needs to be changeable, such as in inventory management.

Advantages
- Changeable: the data on read/write RFID tags can be rewritten, making it changeable.
- High data storage: read/write RFID tags can store a large amount of data.
- Good flexibility: the data on read/write RFID tags can be rewritten multiple times, so they are well suited for applications where the data needs to be updated frequently.

Disadvantages
- High cost: read/write RFID tags are more expensive than read-only and WORM RFID tags.
- Maintenance required: read/write RFID tags need to be regularly replaced or recharged.

Phased Array Antennas

A phased array antenna is an advanced type of antenna system that consists of multiple individual antenna elements. These multiple individual antennas are arranged in an array configuration, a regular grid pattern, which can be controlled electronically to direct the transmitted or received RF signals in a particular direction. The signals from these individual elements are combined and manipulated using a process called phase shifting, which allows the antenna to electronically direct its beam radiation pattern without having to physically move the antenna. Additionally, the phase and amplitude of each individual antenna element can be controlled individually to electronically direct and shape the antenna's radiation pattern.

The ability to control the direction of the radiation pattern swiftly and accurately is one of the key benefits of using a phased array antenna, by adjusting the phase of the signals at each antenna element, the overall beam can be directed in a specific direction or multiple directions simultaneously. This capability to change the beam direction electronically makes phased array antennas very apt for use in radar systems, satellite communications, and wireless communication systems.

The phased array antenna accomplishes this by adjusting the relative phase of the signals fed to or received from each antenna element. This further allows easy manipulation of the constructive and destructive interference patterns of the waves, resulting in a single beam or multiple beams pointed in a specific direction or in multiple directions.

Phased array antennas offer several benefits over traditional mechanically steered antennas. These include:

Faster beam steering: Phased array antennas can alter the direction they are directed instantaneously as the direction of the beam is managed electronically. This allows instant tracking or scanning of an item or an asset or any target.

Increased reliability: Phased array antennas are not susceptible to mechanical failures like traditional mechanically directed antennas as they have no moving parts, offering prolonged operational life.

Simultaneous multiple beams: Phased array antennas can track multiple items simultaneously as they can produce multiple beams at the same time.

Phased Array RFID Antennas

A phased array RFID antenna is an antenna system that coalesces the tenets of a phased array antenna with RFID technology. A phased array RFID antenna includes multiple individual RFID antennas where the phase and amplitude of each individual antenna are controlled to improve the performance of the RFID system accomplished through electronic beam steering, beamforming, and generation of multiple beams (Figure 7.13).

Phased Array RFID Antennas in Healthcare

A phased array RFID antenna can be utilized in healthcare in multiple ways to improve patient care, asset management, or staff tracking to improve the overall efficiency of the healthcare system.

Figure 7.13 Phased array RFID antenna.

Source: impinj.com

Patient Tracking and Identification

A phased array RFID antenna can help locate and identify patients within a hospital facility using RFID Tags attached to patient wristbands. This can help the hospital analyze and reduce patient wait times, ensure that right treatments are delivered to the patients, and reduce the risk of wrong patient identification.

Asset Management

A phased array RFID antenna system can be used to track the location of hospital assets in real time by tagging hospital assets with RFID tags. Such a system can help reduce, if not prevent, loss or theft of assets or other high-value critical items, optimize the utilization of all hospital equipment, and facilitate faster and more efficient tracking or locating of hospital assets.

Staff Tracking

A phased array RFID antenna system can also be used to track the location of nurses, physicians, and other staff members in a hospital, especially at times of emergency or in situations where the staff is experiencing duress. Such a system can also help monitor staff workload and ensure appropriate staff are available in critical areas when needed, which can help improve overall staff management.

RFID Technology in Healthcare

While RFID technology continued to be commercialized in other industries like retail and manufacturing, its adoption in the healthcare industry was not quick, healthcare took some time to understand and realize the benefits of this technology. In the early days, RFID tags were primarily used for inventory purposes, Hospitals would use them to keep track of their medical supplies, and pharmacies would use them to keep track of the drugs. But the healthcare industry is catching up and is finding the RFID technology to be extremely beneficial in multiple areas of the Hospital. The multiple applications of RFID in healthcare include patient tracking, staff management especially in applications like staff duress, asset management for asset tracking and tracing, inventory and PAR level management, medication management, and for managing tissues and samples in Hospital laboratories.

Impact of RFID Technology in Healthcare

RFIDs have started to make some tremendous impact in the healthcare industry as a whole. Some of the ways RFID has been making impacts in healthcare include:

Patient Identification and Safety

RFID technology is helping to improve patient safety in a number of ways, by attaching RFID tags to patient wrist bands at the time of check-in, Hospitals are able to identify patients accurately and quickly. This ensures the patient location is always known so doctors and nurses are able to keep track of their patients throughout the hospital, thereby making sure proper care is always delivered to the patients at the right place and when needed. This information is then used to help reduce the risk of errors, such as wrong-site surgery, wrong medication disbursement, leaving a patient unattended, reducing patient wait times, and improving overall staff communication.

Another way RFID is improving patient safety is through the use of tamper-evident or tamper-resistant RFID tags. These tags provide positive identification of patients, help prevent patient elopement, prevent abduction of infants, and are also used to deliver the correct medication or treatment on time. In addition, RFID tags are also used to monitor vital signs and provide alerts if a patient's condition changes unexpectedly.

Inventory Management

Another important application of RFID in the healthcare industry is inventory management. Hospitals and other healthcare facilities are constantly dealing with the challenge of managing inventory, particularly when it comes to high-value items such as surgical equipment and medications.

RFID can help to automate the inventory management process by providing real-time data on the location and status of various items. This information can be used to generate reports that show where items are located, how many are in stock, and when they need to be replenished. In addition, RFID tags can be used to trigger alerts when items are moved or removed from designated areas.

RFID can be used to manage medical inventory, including, medication, syringes, and IV bags. RFID tags can be placed on these items, and an RFID reader can track their location and usage. This allows hospitals to reduce waste and improve inventory management.

Asset Tracking and Tracing

In a hospital setting, it is often critical to know the whereabouts of certain assets at all times. This includes everything from medical equipment to pharmaceuticals. In the past, this was typically done through manual inventory methods, which were both time-consuming and prone to error.

Enter RFID technology. RFID tags are small devices that can be affixed to assets in order to track their location via radio waves. This way, hospitals can quickly and easily keep track of their assets, without the need for manual inventory methods. Hospitals are large, complex organizations with a vast array of assets, from medical equipment to patient records.

In addition to managing the data on the RFID tags themselves, it is also important to track the location of the assets they are attached to. This can be done using GPS tracking devices, or by manually entering the location information into the database. Either way, it is important to have a system in place so that you can easily locate any assets that are missing or have been moved without authorization.

Data Management

RFID tags contain electronically stored information that can be read by an RFID reader. This data can include the serial number of the asset, as well as other important information such as the date of purchase, or manufacturing date. In order to manage this data effectively, it is important to have a centralized database where all of this information can be stored and accessed. There are many different software programs available that can help with this, and it is important to choose one that will be compatible with your existing systems.

Quality Control in the Healthcare Industry

The healthcare industry is one of the most important industries in the world. It is responsible for the health and well-being of billions of people. As such, it is essential that quality control procedures are in place to ensure that patients receive the best possible care.

RFID technology can play a vital role in quality control in the healthcare industry. By tracking assets and equipment, it is possible to ensure that they are properly maintained and sterilized. This can help to prevent the spread of infection and disease. Additionally, RFID can be used to track medication and supplies, ensuring that they are properly stocked and dispensed.

Monitoring Critical Systems

Another important application of RFID technology in the healthcare industry is monitoring critical systems. This includes systems such as HVAC (heating, ventilation, and air conditioning), electrical, and plumbing. These systems are essential to the operation of healthcare facilities, but they can also be expensive to maintain.

Personnel Management

RFID can be used for personnel tracking in hospital settings. For example, if a staff member is working in a hazardous area, their RFID tag can be used to monitor their location and ensure they are not in any danger. This technology can also be used to track the movements of patients within a hospital. By understanding the flow of patients throughout the facility, hospitals can improve the efficiency of their operations.

Rapid Device Recalls and Sunsetting Standards

The recalls of millions of pacemakers and other devices in recent years have underscored the importance of having comprehensive tracking and tracing capabilities for medical devices. In particular, the ability to quickly locate and remove all affected devices from service is critical to protecting patient safety.

RFID technology has emerged as a key solution for improving device traceability and recall efficiency. By affixing small RFID tags to devices during manufacture, hospitals can track the whereabouts of individual items throughout their lifecycle – from initial delivery and installation, through routine maintenance and repairs, to eventual removal from service.

In addition, RFID-based tracking systems can provide valuable insights into device utilization rates, which can help hospital staff make more informed decisions about repair and replacement planning. As medical devices become increasingly complex and expensive, maximizing their lifespan is a top priority for hospital administrators.

Differentiates Your Assets from Consignment Inventory

One of the main reasons why hospitals and other healthcare facilities have been investing in RFID technology is to keep track of their assets. With an increasing number of consignment inventory, it is becoming difficult to keep track of what belongs to the hospital and what doesn't.

An RFID system can help you manage your consignment inventory better by tracking the location of your assets at all times. This way, you will always know where your equipment is, and you can be sure that it is being used properly.

Loss Prevention

Another advantage of using RFID technology in the healthcare industry is that it can help you prevent losses. For example, if you have a lot of equipment that is constantly being moved around, it is easy for some of it to get lost.

With an RFID system in place, you can track the location of your equipment at all times and know exactly where it is supposed to be. This way, you can quickly locate any missing items and make sure that they are returned to their proper place.

How Does RFID Asset Management Help Hospitals?

Asset management provides full visibility of all equipment, tools, and other assets. It helps in reducing the cost of inventory, maximizing utilization, and improving decision making. All these factors are extremely important for a hospital as it helps in providing better patient care while reducing the overall cost.

Asset Tagging for Check-In Check-Out Process

Tracking the location and status of medical equipment is critical for patient safety and for maintaining an efficient workflow in hospitals. Medical equipment is expensive, and downtime can be costly. With an RFID asset management system in place, nurses and other hospital staff can quickly check-out or check-in assets as needed. The system can also provide alerts if an asset is not returned on time, preventing delays in patient care.

RFID-Enabled Asset Tracking for Preventative Maintenance

Another important benefit of RFID technology is its ability to track assets for preventative maintenance purposes. By tracking when and where an asset is used, hospital staff can proactively schedule maintenance before problems arise. This helps to avoid unplanned downtime and keeps medical equipment working at peak performance.

Use Analytics to Make Effective Decisions

Hospitals generate a large amount of data on a daily basis. Asset management systems can collect this data and use it to generate valuable insights. This information can be used to improve hospital operations, make better decisions about purchasing new equipment, and provide evidence-based support for process changes.

Real-Time Tracking of Medical Devices

In many cases, it is important to know the precise location of medical devices in order to ensure patient safety. For example, if a critical piece of equipment needs to be sterilized, it is important to know exactly where it is and when it will be available. With RFID asset tracking, hospitals can achieve real-time visibility of medical devices, allowing them to make better decisions about patient care.

RFID technology has a wide range of applications in the healthcare industry. From asset management to preventative maintenance, RFID can help hospitals improve their operations and provide better care for their patients. With the right system in place, hospitals can use RFID to track assets, schedule maintenance, and make effective decisions based on data.

How RFIDs Improve Patient Safety and Hospital Workflow

Patient safety and hospital workflow are two areas where RFID technology can have a major impact. By tracking patients, medical staff, and equipment with RFID tags, hospitals can reduce the risk of errors and improve efficiency.

RFID tags are usually affixed to wristbands or ID badges worn by patients and staff. The tags contain unique IDs that can be read by RFID readers. This information can then be used to track the location of patients and staff, as well as hospital equipment.

Enhancing Infant Safety

One of the most important applications of RFID in healthcare is enhancing infant safety. In hospitals, it's not uncommon for babies to be mixed up with other infants who have similar physical characteristics. This can lead to devastating consequences, such as a baby being given the wrong medication or undergoing the wrong surgery.

RFID tags can help prevent these mix-ups by providing a unique ID for each infant. The ID can be linked to the infant's medical records, so that staff can quickly and easily identify the correct patient. RFID tags can help prevent these mix-ups by providing a unique ID for each infant. By tracking infants with RFID tags, hospital staff can quickly and easily identify the correct baby when it's time for them to be discharged.

Improving Hospital Workflow

Hospitals are complex organizations, with many moving parts. RFID technology can help streamline hospital workflow by reducing the need for manual tasks, such as searching for patients or equipment.

For example, imagine that a patient needs to be moved to a different room. Rather than manually searching for the patient's file, staff can simply use an RFID reader to locate the patient's wristband or ID badge. This information can then be used to update the patient's records and ensure that they are transferred to the correct location.

RFID technology can also be used to track hospital equipment, such as wheelchairs and IV pumps. This information can help reduce the time needed to find and retrieve equipment. In addition, it can help hospitals keep a better inventory of their medical supplies.

Reducing Medication Errors

Medication errors are a major problem in healthcare. In fact, it's estimated that one in every five hospital patients is harmed by a medication error.

RFID tags can help reduce the risk of medication errors by providing a unique ID for each patient. This information can be linked to the patient's medical records, so that staff can quickly and easily identify the correct patient when it's time to dispense medication.

RFID technology can help reduce the risk of medication errors by tracking medications throughout the hospital. For example, many hospitals now use RFID-enabled cabinets to store and dispense medications. The cabinets track which medications have been taken and when they were taken. This information can then be used to ensure that patients receive the correct medication at the correct time.

In addition, some hospitals are using RFID tags to track individual pills. These tags contain information about the medication, such as the dosage and route of administration. By scanning a pill's RFID tag, staff can quickly and easily verify that the correct medication is being dispensed.

Using Tags to Monitor Temperature for Supply Storage

Another important application of RFID in healthcare is temperature monitoring. Many medical supplies, such as vaccines and blood products, must be stored at specific temperatures to remain effective.

RFID tags can help hospitals track the temperature of their medical supplies. The tags contain sensors that measure the temperature of the surrounding environment. This information can then be used to ensure that medical supplies are being stored at the correct temperature.

In addition, some RFID tags contain GPS chips that allow them to be tracked in real time. This information can be used to monitor the location of medical supplies and ensure that they are not being exposed to extreme temperatures.

How to Choose the Right RFID?

Choosing the right RFID technology for your healthcare organization can be daunting. With all of the different types of RFID tags, readers, and antennas available, it's hard to know where to start. So, how do you choose the right RFID technology for your needs?

Here are a few things to consider when choosing RFID technology for your healthcare organization:

What Are You Trying to Track and Trace?

This is the most important question to answer when choosing RFID technology. Are you trying to track medical equipment, pharmaceuticals, or both? Each type of asset has different tracking and tracing needs. For example, medical equipment is often larger and has a longer lifespan than pharmaceuticals. As a result, you'll need a different type of RFID tag for each type of asset.

What Is Your Budget?

RFID technology can be expensive. You'll need to factor in the cost of tags, readers, antennas, and software when choosing RFID technology for your healthcare organization. Make sure to create a realistic budget that takes into account all of the costs associated with RFID technology.

What Are Your Space Constraints?

Consider the size of your facility when choosing RFID technology. If you have a large facility, you'll need a robust RFID system that can cover a wide area. If you have a small facility, you may be able to get away with a less expensive RFID system.

What Is the Level of Security Needed?

Healthcare organizations handle sensitive information on a daily basis. As such, it's important to choose an RFID system that meets your security needs. Make sure to consider who will have access to the RFID system and what type of information they will be able to view.

Do You Need Global Tracking or Just Tracking within Your Facility?

If you need to track assets globally, you'll need an RFID system that can handle global tracking. This type of system is typically more expensive than a system that only tracks assets within your facility.

Polarization

There are two types of RFID tags – linear polarized and circular polarized. Linear polarized RFID tags are more expensive, but they offer a longer read range. Circular polarized RFID tags are less expensive, but they have a shorter read range. Choose the type of RFID tag based on your needs.

Gain

The gain of an RFID tag is the amount of power that is radiated from the tag. The higher the gain, the longer the read range. Choose an RFID tag with a high gain if you need a long-read range.

These are just a few things to consider when choosing RFID technology for your healthcare organization. By taking the time to answer these questions, you'll be able to choose the right RFID system for your needs.

RFID technology has the potential to revolutionize healthcare by making information more accessible and improving patient care. The use of RFID tags can help reduce medical errors, speed up processes, and improve communication between staff members and patients. While there are some concerns that need to be addressed, such as privacy and security, the benefits of RFID technology in healthcare far outweigh any possible risks. Healthcare providers should begin exploring how they can implement RFID technology in their facilities to improve patient care.

References

[1] L. Scanlon. (2003). Good Vibrations. Electronic music's Soviet roots. Retrieved December 25, 2022, from MIT Technology Review Online: https://www.technologyreview.com/2003/03/01/234404/good-vibrations-4/

[2] L. A. Hyland, A. H. Taylor, and L. C. Young (27 Nov. 1934). System for detecting objects by radio, U.S. Patent No. 1981884,

[3] April 1935: British Patent for Radar System for Air Defense Granted to Robert Watson-Watt. American Physical Society. Retrieved December 25, 2022, from APS News, April 2006, (Volume 15, Number 4) Online: https://www.aps.org/publications/apsnews/200604/history.cfm

[4] G. D. Romagnosi. (2009). In Encyclopedia Britannica. Retrieved October 25, 2009, from Encyclopedia Britannica Online: http://www.britannica.com/EBchecked/topic/507231/Gian-Domenico-Romagnosi

[5] L. Bowden (October 1985). The story of IFF (Identification Friend or Foe)." IEEE. Retrieved December 25, 2022 from IEEE Proceedings, Part A: Physical Science, Measurement and Instrumentation, Management and Education – Reviews, Volume 132 (6), pp. 435–437. ISSN 0143-702X.

[6] J. Landt. (2001) Shrouds of Time: The History of RFID. Association for Automatic Identification and Mobility: Retrieved December 25, 2022. Online http://www.aimglobal.org/technologies/rfid/resources/shrouds of time.pdf
[7] H. Stockman (1948). Communication by means of reflected power, Proc. IRE pp. 1196–1204.
[8] C. A. Walton (August 14, 1973). U.S. Patent 3,752,960.

Chapter 8

Developing an Institutional RFID–RTLS Strategy and Management Plan

Paul H. Frisch

Introduction

The changing healthcare environment has driven hospitals to critically evaluate and optimize their operations to enhance patient treatment and care while focusing on financial constraints. The hospitals have moved to support an increasing outpatient care environment, driving increasing in-patient acuity levels. At many institutions, these changes have been accompanied with regulatory and financial constraints impacting the operating budgets. Specific initiatives, such as the Joint Commission's patient safety goals, electronic medical record, and meaning full use, have driven institutions to redesign their operations and processes, using technology solutions to achieve these goals, while optimizing operational workflow and resource utilization.

To deal with the complexity of patient care workflows, enhance patient diagnosis, treatment, care, safety, and satisfaction, the design of Intelligent Health Systems has focused on the integration of diverse technologies, to provide a seamless exchange of information and optimize workflows and operations. This has driven hospitals to integrate location-based technologies including RTLS and other RFID and BLE solutions into a variety of hospital operations and processes.

DOI: 10.4324/9781032690315-8

The overall cost of an RTLS/RFID deployment not only includes the capital purchase but more significantly the on-going operational cost. RTLS/RFID systems represent a significant investment where the return on investment (ROI) is realized through the deployment of a broad set of applications and use cases. Location technologies include a broad base of solutions, including dynamic institutional visualization and tracking using localized choke point solutions, workflow monitoring and metrics, process verification and validations, and alert notifications.

This chapter outlines the considerations and strategies for the deployment of large-scale RTLS/RFID solutions. These strategies include system scaling, requirements for accuracy and resolution, program management, managing an increasing quantity of tags and tag types, monitoring the effectiveness of the solution, managing impacts of complexity and interference, and establishing a dedicated support model and structure. The paper highlights multiple use cases or applications targeting enhanced asset management, optimization of workflow, and process validation.

Over the past decade, healthcare institutions have experienced an increase in patient acuity levels concurrent with an increase in the number of critical care monitored beds, medical devices, emergency room visits, and post-surgical cases [1–3]. With the advent of diminishing family practices, many patients use hospital emergency departments as their primary method of general healthcare impacting patient flow, throughput, and volume. In parallel, hospitals are implementing a broad base of clinical data systems, including Electronic Medical Record (EMR), alarm management systems, and medication administration platforms which present new data sets with an increasing volume of information presented to the clinical staff [4,5]. This combination of issues directly impacts the care providers workload, requiring them to manage an increased number of patients while processing larger volumes of information (big data).

The consequence of these factors contributes to the reducing direct patient-to-staff time, increasing number of errors, which include medication errors [6,7], device and systems operating errors, and errors associated with patient identification. These trends have raised concern within regulatory agencies and hospital accreditation organizations (Joint Commission), to establish guidelines to address these patient safety concerns [8]. These patient safety guidelines have targeted the need for improved asset management and process verification, such as hand hygiene and cleaning processes.

These processes have been positively impacted through the utilization of RFID. Many institutions have deployed RFID, to enhance decision support, optimize workflow, validate clinical and business processes, and improve patient throughput, supply management, and sample collection. Active RFID solutions are used for dynamic and real-time institutional-level tracking of assets whether it be devices, inventory, staff, or patients. Passive RFID solutions have focused on supply chain and localized inventory management projects.

The objective of an institutional RFID/RTLS solution is to provide, unique identification and global visualization focused on the tracking of institutional assets to optimize clinical workflows and business processes. Optimized workflows commonly target regulatory compliance, validation of processes, and patient safety by minimizing search times increasing time and staff availability for patient care.

Strategic Planning and Design

Design of an institutional RFID/RTLS system requires careful planning, input from each of the stakeholders, understanding of the specific use case objectives and the environmental factors that can directly impact RFID effectiveness, including wireless infrastructure, localization accuracy, and potential interference. Without a well-organized plan coupled with comprehensive design and investigative studies and pilot solutions the plans, deliverables and objectives can fail to meet the expectations of the institution, for several reasons highlighted below:

1. Design Issues based on scalability and accuracy
2. Unreliable performance
   a. Poor tag management, association, and mapping
   b. Issues associated with wireless infrastructure
   c. Unexpected interference
3. Inability to demonstrate performance objectives, metrics, or ROI

A well-designed institutional solution addressing a broad set of applications implementing a broad set of use cases requires scalability enabling the system to define the specific accuracy of locating specific assets, whether it be equipment, staff, supplies, or pharmaceuticals as a function of the specific requirement of the use case. Hospitals will deploy many types of identification and locating technologies, RTLS, RFID, active and passive, BLE and most recently 5 G, supporting many use cases with varying requirements for range, resolution, and location accuracy to support the needs of inventory tracking applications and workflow metrics to enable optimization. In many cases, a minimum of room-level granularity will be necessary to achieve the appropriate process metrics; and in the case of supply chain management applications, cabinet or shelf level granularity is commonly required. As the need for greater resolutions increases additional dedicated infrastructure to localize asset position will be required. This can include low-frequency exciters, IR or ultrasound transmitters, and receivers. This additional infrastructure significantly drives up the cost of the RFID implementation as well as the support and maintenance requirements. It is important to reiterate that scaling the accuracy and resolution needs to the specific use case enables and institution to minimize systems cost by limiting unnecessary resolution based on the use case application. As an example, validating inventory presence can be achieved through Wi-Fi access point resolution only, while the location of specific assets may require minimum room-level accuracy. This mixed-mode design can significantly reduce the overall institutional expense.

Return on Investment (ROI)

RTLS/RFID should be considered an institutional solution and represents a significant investment for the institution and can represent a potential for significant ROI, both financially, in terms of patient safety through the optimization of business processes and clinical workflows. As previously highlighted, the total institutional ROI is based on the deployment of many applications and use cases including combinations of asset management and workflow, each contributing a portion of the overall ROI. Without the planned deployment which exploits a board set of use cases, it becomes difficult if not impossible to achieve the ROI projections.

Leveraging of Existing Infrastructure and Alternative Technologies

RFID systems can require a dedicated infrastructure consisting of a network of transmitters and receivers to be used to provide location. Alternatively, organizations supporting a wireless infrastructure, such as 802.11 and or a robust 5 G platform can leverage the existing infrastructure to provide cost savings. The determination of the best design is a function of many considerations, including the type and number of tags required and whether the applications are localized or global. At Memorial Sloan Kettering Cancer Center (MSK) prior to deployment, several critical analyses and pilot studies were performed to evaluate the best options as a function of clinical environment and application objectives. It was determined that the use of the existing wireless infrastructure represented the ideal starting point for our deployments. Additional supporting hardware (LF and ultrasound transmitters) were added to areas requiring increased resolution. It also became clear that both active and passive technologies would be used in combination along with identification data from our legacy technologies (barcodes and IR).

Design and Pre-Deployment Analysis, Evaluation, and Pilot Studies

To ensure that the system design will achieve the expected result, data quality and outputs from the RFID applications must be verified based on several pre-deployment analyses, evaluations, and pilot studies. These studies are used to identify any unexpected issues resulting from bandwidth issues, network coverage, reliability, and unexpected interference. The results from these studies provided the design specifications and considerations and validated our ability to achieve the design and information objectives.

Network Impact

Deploying active RFID tags that transmit over the wireless infrastructure requires an analysis of the frequency of transmission and the duration of the signal contributing to the calculation of network traffic. While the information transmitted from an individual tag is very small (typically several hundred Kbytes), the combined effect of thousands or tens of thousands of tags simultaneously transmitting can potentially have some impact on the network.

As an example, to assess the impact on network traffic, Biomedical Engineering in collaboration with Information Systems performed an analysis that examined a worst-case scenario of 1000 active tags within the range of single access point broadcasting 156 Kbytes at 8-second intervals. The over-air and network (LAN) impact indicated that within this extreme scenario only 0.03% of a 100 base T wired network would be utilized for RFID applications indicating a limited potential for RFID impact on hospitals network bandwidth.

Interference

MSK additionally studied the potential electromagnetic interference induced by the RFID on medical devices as well as potential impacts on RTLS accuracy from other devices [9]. Electromagnetic energy-generating devices, such as RFID, have the potential to interfere with other electronic devices. Within the patient care environment, any interference with medical devices, providing life support, treatment, or diagnostics can directly impact patient safety and introduce risk. Interference is directly related to the energy transmitted, the relative distance, and geometry between the medical devices and transmission source. All hospitals are required to assess potential patient risks. At MSK, Biomedical Engineering has the responsibility to perform its due diligence to assess and minimize the potential risk resulting from device failures. Our internal test analysis performed and placed six Wi-Fi RFID tags on each of the surfaces of the medical devices, including physiological monitoring, infusion pumps, and ECG machines. Tags were programmed to transmit at 1-second intervals. Device performance was tested and verified with an augmented preventative maintenance (PM) test procedure. No operational impact was observed during system operation, archived data, connectivity, or data output from the devices.

Collaborations with the FDA enabled further expanded and controlled experiments to study RFID-induced impact on medical devices at the FDA White Oaks, EMI test facility. Devices tested were provided by MSK and again included monitoring devices, ECG systems, and infusion pumps. The impact of observed EMI on medical devices was classified either as clinically significant (Class I) or not clinically significant (Class II). As expected, the results documented potential device susceptibility to RFID-induced EMI as a function of frequency and power [10].

RFID Clinical Application Pilot Studies

Prior to final product selection and deployment, Biomedical Engineering performed a series of tests and operational pilots for approximately 6 months to evaluate RFID capabilities and validate RFID performance, data consistency, and success on the projected use cases. MSK configured a 42-bed inpatient unit tagging patients, staff, and devices. Each bed and room were outfitted with supporting localization hardware (LF exciters) to enable bed-level localization. RFID performance metrics were extracted from the system to confirm RFID accuracy and resolution for both triangulation and localized ranging. The pilot demonstrated the effectiveness of RTLS/RFID to meet the objectives and enabled the projection of ROI focused on enhancing asset/device management. The pilot data coupled with the projected ROI enabled Biomedical Engineering to justify the purchase of the system and start the deployment of the proposed user cases.

III: RFID Program: Applications (Use Cases)

Since RFID represents an institutional solution, specific use cases commonly cross departmental boundaries requiring a coordinated approach to define the use case, set the objectives, and define the deliverables. In addition, managing an expanding RFID implementation requires a coordinated and centralized approach to plan, prioritize, and manage all aspects of deployment. As you may expect once a scalable RFID solution is available to the institution, hundreds of projects are proposed and requested. It is important that the organization review all the proposals, evaluate the projected ROI, benefit to patient management or safety, and prioritize the projects as a function of institutional importance and budget. Creation of a leadership group (steering committee) establishes a consistent project review, budgeting, and oversight. The committee reviews requests for new or expanded RFID project proposals and prioritizes the projects providing a pathway for funding. The committee functions to monitor on-going projects and prioritize future deployments ensuring that the objectives are successfully met and continue to provide the ROI projected.

Within an Intelligent Healthcare System, a combination of passive and active RFID/RTLS solutions are projected to be deployed to enable identification, comprehensive device tracking, and the validation of critical processes. One approach is that all institutional medical devices are tagged and identified with passive RFID assigned at the time of incoming inspection and inventory. Select devices, such as mobile devices and patient critical devices additionally have active tags, beaconing at set intervals which can include RFID and or BLE technologies. In most hospitals, there continues to be a reliance on legacy bar code applications, which identify assets, which must be considered in the overall institutional strategy and system design, as illustrated in Figure 8.1.

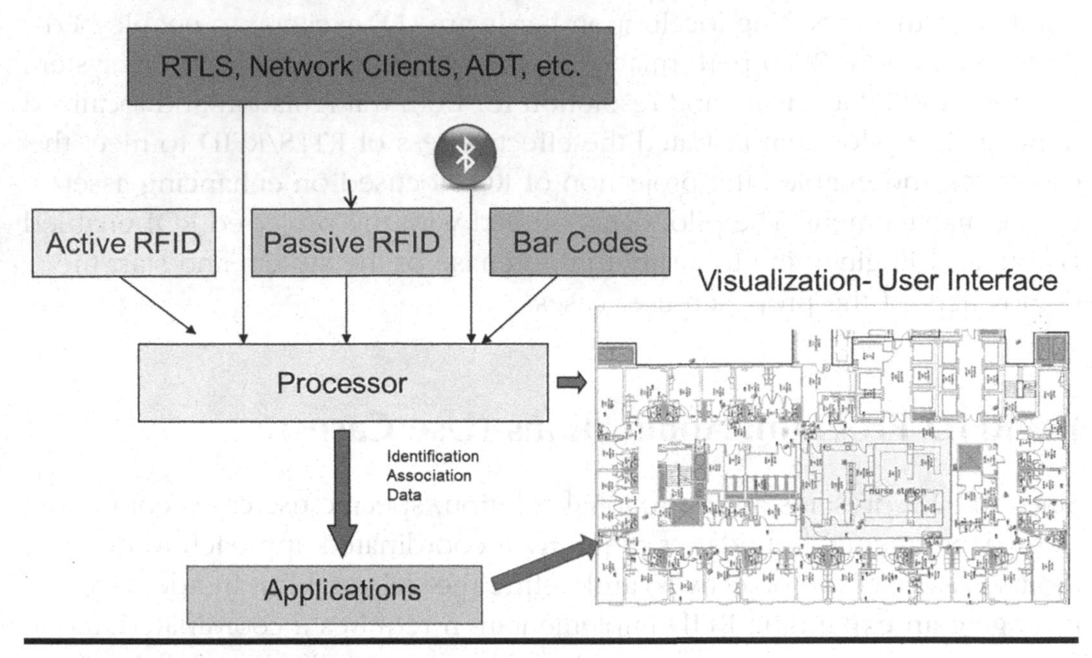

Figure 8.1 Institutional RFID implementation.

Use Cases

All institutions depend on the ability to uniquely identify and potentially track these assets for a variety of applications and usc cases. These use cases cover a broad spectrum of needs in an Intelligent Health System, which includes audit, financial management, regulatory compliance, critical device recovery, and association of devices to patients and processes. The following examples demonstrate how specific use cases can directly impact hospital operations, enhance patient care and safety patient, and impact the satisfaction of the patient experience.

Use Case: Device Tracking – Grant Audit Device Verification

Academic institutions that engage in grant-funded projects represent a very specific asset tracking target for RFID which also can provide a significant ROI, targeting regulatory audit and compliance. These Institutions typically operate research programs functioning in-part via grant funding from a variety of Foundations and Government agencies, such as the NSF and NIH. These grants can include the funding for specific equipment and devices necessary to perform the specific research and investigations outlined within the grant. As part of the grant process, grant-issuing agencies audit these institutions to ensure the proper use of grant funds. Depending on the number of grants and specific agencies, audits can occur 1–2 times per year. As part of the audit process, institutions are required to confirm and demonstrate the presence of any devices purchased and funded under grant funding. Prior to RFID, this audit process was typically performed by outside consulting companies visually validating the devices/inventory, providing documentation to the auditing agencies at a very significant cost per audit.

Deploying an active Wi-Fi triangulation-based RFID tracking solution utilizing the hospital's existing wireless infrastructure provides a unique method of identifying the devices and providing the real-time location of all grant-funded or purchased devices throughout the institution. This capability eliminates the need for any external validation services and enables the inventory to be identified and located at any moment in time. The savings from the first 1–2 audits essentially can cover the cost of the initial base RFID investment.

More importantly, the implementation of the Grant Audit Use Case, essentially makes institutions RFID capable enabling a Wi-Fi triangulation RFID solution to be available and deployed at all locations within eh health system supporting Wi-Fi capability.

Use Case: Regulatory Inventory Validation and PM Compliance

The next series of deployed use cases focused on regulatory compliance and asset management. All hospitals are required to demonstrate regulatory compliance based on the guidelines set by the Centers for Medicare & Medicare, DOH, Joint Commission, and other state and local agencies. The primary regulatory requirements for medical device management focus on ensuring equipment safety and operational integrity. This requires Biomedical Engineering departments to maintain a highly accurate medical device inventory, coupled with a robust PM program to maintain the required compliance. New guidelines set high standards for PM compliance of critical devices at 100%, with non-critical devices at a minimum of 95%.

All devices support passive RFID tags, which enables staff to perform a simple walkthrough and using a hand-held RFID reader scanning all the devices at predetermined intervals (typically monthly) which is then co-ordinated with the preventative maintenance schedule. The unique RFID identifiers are downloaded into the device database, compiled, and stored. The inventory database automatically filters devices that are read on multiple scans and compares the newly scanned devices to the archived existing inventory. The variance between the two lists quickly identifies devices that are no longer RFID visible as well as devices that could not be located for PM. Subsequent PMs are performed on devices where possible maximizing PM compliance. Devices that are no longer RFID visible are further searched within the support from the departments. Those that are confirmed as missing or unknown are defined as in-active increasing the accuracy of the inventory. If in-active devices become RFID visible on future scans or appear for service and maintenance, devices are dynamically restored to the active inventory and subject to PM requirements. There are issues with devices stored and potentially shielded in metal cabinets and behind highly reflective surfaces such as glass. This does require Biomedical Engineering to be more diligent and obtain access to these areas. However, the RFID process enables the inventory to be regularly and dynamically monitored to maximize inventory accuracy and identifies or locates devices that are PM due in cabinets, draws, or not easily visible, enhancing the overall rate of PM compliance. This process does enhance the device decommissioning process improving positive confirmation of devices returned to the manufacturer, sold, or discarded.

Since not all hand scanning can be performed simultaneously in all locations, mobile devices can be missed using this process exclusively. As previously noted, mobile devices are additionally tagged with active Wi-Fi RFID tags enabling real-time tracking of mobile devices. By combining the passive and active methods, inventory accuracy and PM compliance are optimized. In parallel, this process can be applied to 3rd Party owned devices, loaners, rentals, etc. to enhance the management of these difficult-to-locate and track devices.

Use Case: Regulatory: Critical Device Tracking and Availability

Fundamental to enhancing patient care is optimizing the availability of resources necessary to provide required patient care. A common problem in most hospitals is locating the devices needed and available for patient care. Ideally, staff minimizes the time necessary to locate devices and resources necessary to manage patients. Capitalizing on the success of active RFID Wi-

Fi-based triangulation can be expanded to a broad set of medical devices commonly difficult to locate at the time of need. These included CMS-classified devices, such as radiological devices, laser systems, and new technologies devices as well as highly mobile devices which could be located anywhere on the patient floor or other locations within the center. The following medical devices are the prime targets for real-time device tracking.

- Defibrillators and AED
- Vital Signs Monitors
- ECG Carts
- Sequential Compression Devices
- Infusion Pumps
- C-Arms Imaging Devices
- Feeding Pumps Anesthesia Systems
- Ventilators
- Defibrillators
- Stretchers, Wheelchairs, and Food Carts

In addition to providing increased access, the RFID tracking of these highly mobile and critical devices enables PM compliance to be maintained at 100%

Use Case: Optimizing Patient Care Identifying Available Infusion Systems

A common issue in hospitals is locating IV infusion systems needed for patient care. However, simply locating devices using RFID alone does not always ensure a successful search outcome. Ideally, knowing the status of the device, in use or not, defines the true availability. One approach is that all infusion devices are tagged with active Wi-Fi RFID tags utilizing event buttons to enable staff to set devices as in-use or available (Figure 8.2).

However, device status is dependent on user compliance and is subject to significant errors. Users were commonly interrupted by urgent patient care issues which resulted in incorrectly setting the state of the devices. These inaccuracies significantly impacted the search and recovery process.

To resolve this issue, Biomedical Engineering in collaboration with the manufacture of the infusion system and RFID system interfaced the infusion systems server directly to the RFID system. For each pump deployed, pump ID and associated status (infusing or not) were sent to the RFID system to

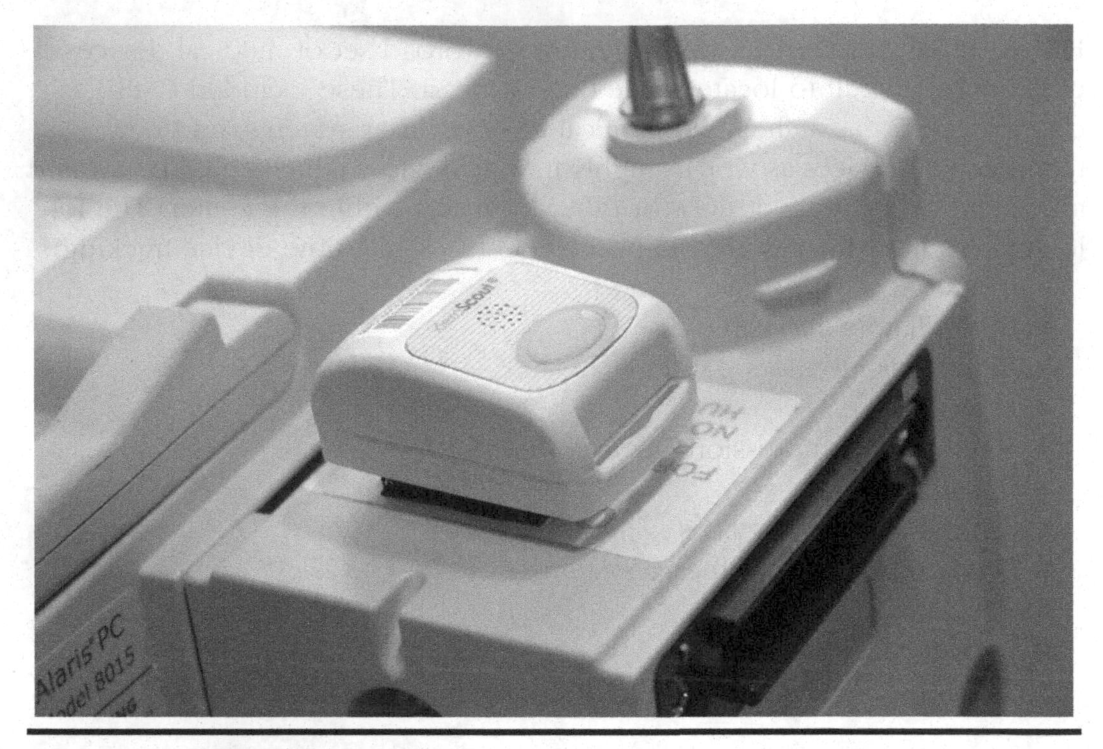

Figure 8.2 RFID event/status button.

uniquely define the state of each device. Clinical staff could easily locate and access the RFID location floor map and visually identify devices available for use on the unit or anywhere in the hospital. This targeted search capability optimizes the search time and ensures the availability of the located pumps. This collaboration between MSK and device manufacturers has resulted in an FDA-approved product commercially marketed by both the Infusion System and RFID manufacturers.

Use Case: Device Optimization – Infusion System Utilization

Once the infusion devices can be identified as available or in-use, statistical data on the infusion pump inventory could be computed. It was now possible to determine and monitor the utilization of the infusion pumps within the overall hospital or down to the unit level. These analytics are used to optimize infusion systems par-level deployment on the patient units, as well as predict the need for additional pumps as the institution expands.

Hour-by-hour changes in infusion pump utilization can be monitored providing precise information of the status and needs of each unit (Figure 8.3). Review of the institutional utilization enables dynamic

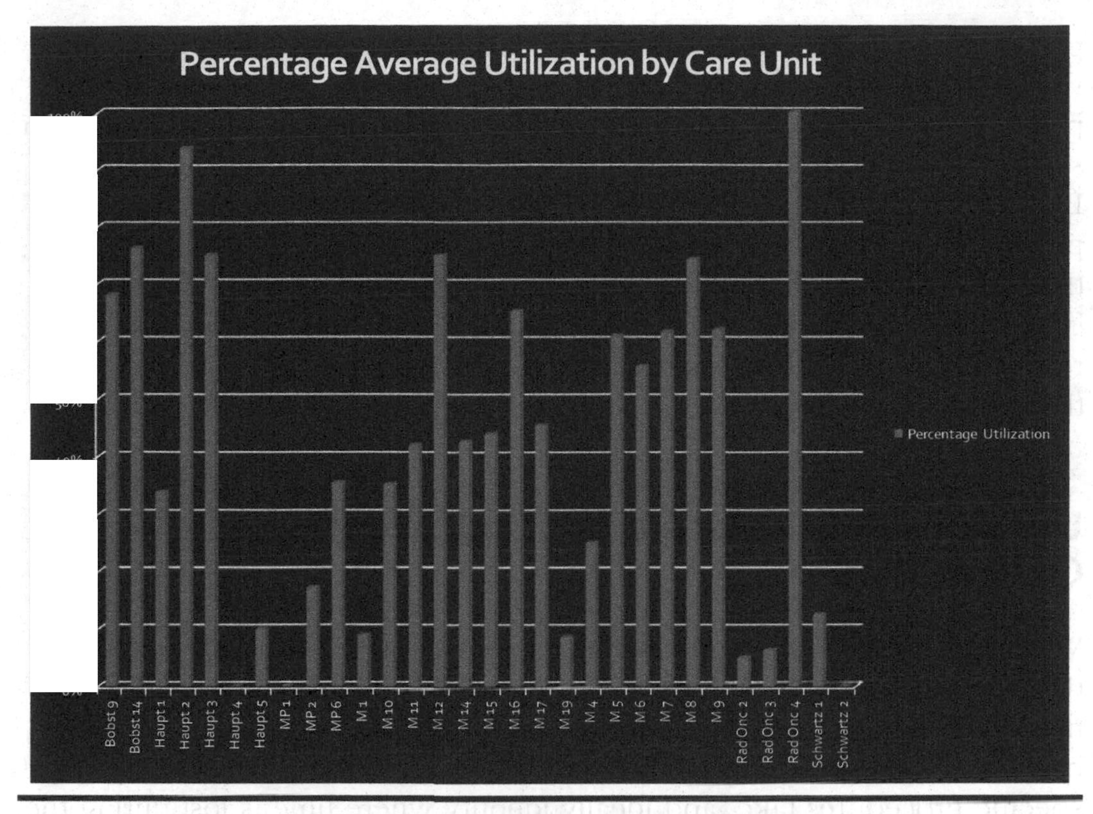

Figure 8.3 Dynamic infusion system utilization by care unit.

adjustment of the PAR levels and redistribution of the inventory to meet the needs of the institution. This dynamic utilization information enables planning and optimization of the inventory to provide maximum availability of medical devices while maintaining as minimal an inventory as possible. Decisions to rent or purchase additional devices are made based on an institutional visualization of device utilization rather than the specific needs of individual areas or departments in the hospital.

Use Case: Patient Safety – Validating IV Formulary

The above integration of the RFID and Infusion Systems additionally enables the deployment of a pharmaceutical formulary validation process critical to patient safety. In addition to extracting the infusion state of the device, the version of the infusion formulary was also available. Pumps are routinely upgraded with new formulary versions. It is critical that all infusion pumps always reflect the most current formulary version. Any discrepancies in the formulary can result in infusion errors and represent a potential safety hazard to the patient. Since the formulary is downloaded wirelessly there is a potential for a formulary download to be missed. The pump may be out of service, in a wireless dead zone, or in transit between sites at the time of download. In collaboration with Pharmacy and Nursing, Biomedical Engineering monitors the formulary version of all infusion pumps throughout the MSK network and sites. Any pump identified with the incorrect formulary is flagged and immediately removed from service based on the known location. The pump formulary is upgraded and restored to service. This process ensures that all devices are maintained in the correct formulary version and enhances patient safety.

Use Case: Patient Satisfaction and Safety Workflow Optimization

Workflow optimization and validation represent one of the major applications for RFID/RTLS solutions. RTLS provides visibility and detailed information about the processes associated with patient care. RTLS enables a hospital to monitor how long patients are in the waiting room, how long specific procedures take, and ideally identify where time is lost during the patient care process enabling enhancements and optimizations to be implemented.

However, ROI associated with workflow can be a difficult metric to present. Unlike asset management where quantitative cost saving can be computed by reducing waste and optimizing purchasing decisions, workflow is a qualitative measure typically associated with optimizations and reductions in workflow process time. These optimizations are frequently reflected as increased staff-to-patient time, which indirectly enhances patient safety, satisfaction, and outcomes. Though reductions in staff hours are used as a benchmark for improvement, rarely does this include staff reductions and changes in the cost of operations. As illustrated in Figure 8.4, the value of RFID focuses on providing real-time metrics of the patient care process for optimizing clinical and business processes.

As before, pilot studies confirmed the concept's feasibility and demonstrated the data metrics. In the pilot, MSK outpatient areas were instrumented with room-level choke point localization hardware (ultrasound transmitters) uniquely defining clinical spaces, such as treatment and exam rooms, and waiting areas. The ability to uniquely localize patient and staff positions enabled the collection of clear metrics for analysis. Table 8.1

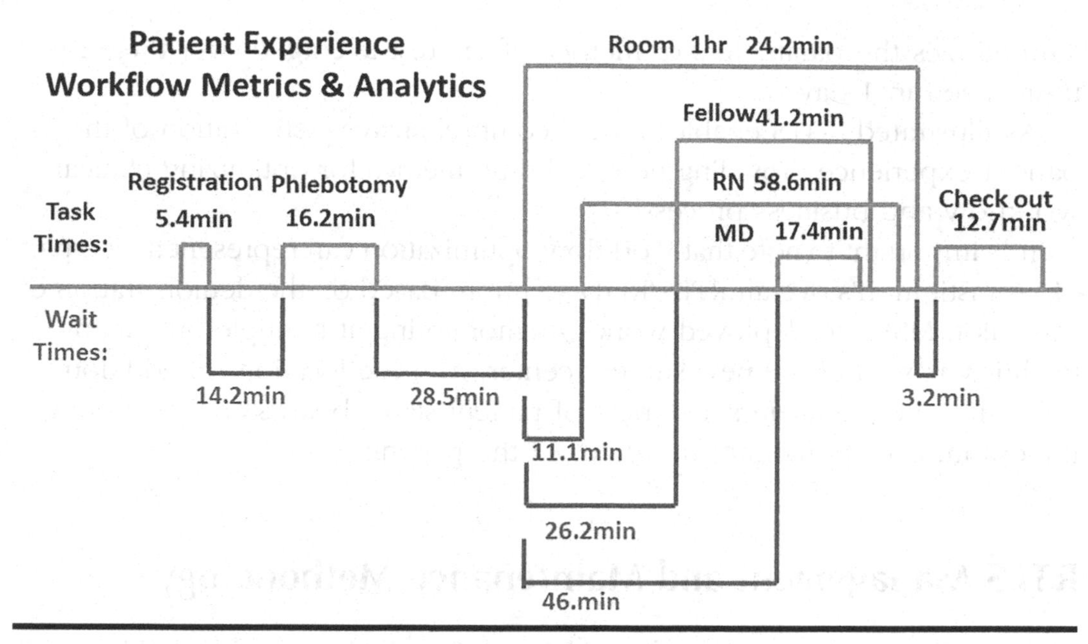

Figure 8.4 The patient experience – workflow metrics.

Table 8.1 Measurement Criteria with Average Data for n = 32 Patients in One Day during Pilot

Measurement Criteria	*Average hr:mm:ss*
Total patient experience time	2:00:17
Time spent at check-in	0:01:34
Wait time for phlebotomy	0:11:46
Phlebotomy treatment time	0:06:33
Wait time to enter an exam room	0:36:00
Total time spent in exam room	0:58:29
Wait time in exam room for clinician (nurse)	0:14:01
Wait time in exam room for clinician (doctor)	0:25:53
Time in exam room with clinician (nurse)	0:16:19
Time in exam room with clinician (doctor)	0:19:10
Total time spent alone in exam room	0:24:35
Wait for check-out	0:06:53
Time spent at check out	0:06:55

summarizes the parameters or metrics of interest averaged over the 32 patients seen in 1 day.

As illustrated, RTLS enables a new comprehensive visualization of the patient experience providing new real-time metrics for optimizing clinical workflow and business processes.

It is important to note that workflow optimization can represent a vital part of an institution's overall RTLS/RFID solution. Based on the demonstration of this pilot, MSK has deployed workflow monitoring at multiple outpatient facilities as well as the new surgery center. At these locations in addition to collecting the parameters, a variety of patient status boards enable waiting patient families to monitor the status of the patient.

RTLS Management and Maintenance Methodology

The continuing effectiveness of any RTLS/RFID system to provide the designed functionality and expected metrics is dependent on the continued management and support of the system, supporting hardware, tags, and reporting tools. Unfortunately, RFID system management, support, and maintenance are often the last consideration in the design of an RTLS/RFID solution. Deployments made without a robust maintenance program are quickly compromised and fail to meet the design objectives and expectations. System support is not a trivial issue and becomes increasingly complex as the number of use cases and tags deployed increases. To maintain a high confidence in location data integrity, it is vital that an institution-wide approach be implemented providing a centralized support and overall management program.

A comprehensive support program focuses on several specific tasks:

- Maintaining and Validating Application Integrity
  - Managing tag to asset associations (patient, staff, devices, and supply)
  - Management of RFID database and integration to other institutional databases (i.e., asset database)
  - Tag Management, Replacement, and Upgrades
    - Tag obsolescence
    - Damaged or lost tags
    - Battery replacement program
- New Application/Use Case Review and Development
- EMI Monitoring

Application Integrity

Consistent RTLS/RFID operation is based on the on-going management and review of the database maintaining the relationship/associations between and RFID tag and the asset, whether patient, staff, supplies, or device. On the device side, the assets are uniquely identified within the institutions asset or inventory database, typically the Biomed CMMS, which integrates with the RTLS/RFID system database tables. Medical devices are commonly re-assigned, sent out for repair, or retired from operation with these changes in status reflected in the Biomed CMMS or asset database. It is vital that these changes are dynamically reflected in the system database or data tables. Without the correct device to tag associations, the accuracy and operational performance are directly impacted. At MSK, Biomedical Engineering has synchronized the inventory database with the RFID system to ensuring that the RFID system tag/association always reflects the status of the devices.

Tag Management

All RFID tags are subject to loss, damage, and obsolescence, requiring a dynamic tag management program to monitor the tag status (date deployed, model, last firmware upgrade, battery changes, lost, exchanged, etc.). As the number of deployed tags moves into the thousands, tag management becomes a complex issue requiring continuous oversight and review. The asset inventory database was expanded to additionally inventory all active RFID tags. These tags were identified by the project/use case, ownership, tag type, software/firmware version, last battery update, projected obsolescence, and tag status (i.e., active, associated, in storage, out of service, replaced). Each tag has a service history associated with the tag detailing changes in the state of the tag.

Battery Change Protocol

Like all battery-powered medical devices, batteries will require replacement at regular predefined intervals to ensure battery replacement in advance of failure. The RFID tag database tracks and schedules battery replacement. As the number of tags increases, it becomes increasingly difficult and eventually impossible to change batteries on the fly based on the RFID systems indication of a low battery condition. Biomedical Engineering has studied battery life as a function of tag chirp rates and has modeled battery replacement based on the expected life of the battery. Results from these experiments have established the projected battery life and enabled us to define the battery change cycle for each tag and application. New generation tags, where chirp or RFID transmission is activated associated with motion, have significantly extended the projected battery life. In these applications, battery status is monitored with battery replacement performed when batteries reach 20% of their full capacity.

Application Development and Reporting

Implementation of new RTLS/RFID applications can represent a large ongoing cost for systems expansion, configuration, application development, visualization, and reporting. Contracting with a vendor to support this function is very limiting and costly. It is recommended that an internal resource knowledgeable in the RFID system configuration, programming, and reporting the status of potential of specific applications and use cases be established providing a dynamic internal resource to the institution. This approach enables rapid use case deployment coupled with the reporting necessary to provide the metrics of interest.

Conclusions

The changing healthcare environment has driven hospitals to utilize RFID technologies to enhance workflows and optimize their business operations. RTLS/RFID has provided new metrics to better understand the complexities of patient care workflows and enables hospitals to enhance patient care, safety, and satisfaction. A well-designed and managed RFID system integrating with the key clinical applications and systems has demonstrated continuous benefit and ROI within the new Intelligent Hospital. The deployment of RFID is not only dependent on the purchase of a system but also on the long-term strategy, planning, and collaboration between all departments to define and prioritize the applications and business expectations. Implementation of a centralized group to develop and implement use cases as well as manage the daily issues and operations is vital to the success of an institutional RFID program.

References

[1] DeGroot H., Patient classification system evaluation, Part I: essential system elements. *J Nurs Adm* 1989; 19(6):30–35.

[2] Van Slyck A., Johnson K. R., Using patient acuity data to manage patient care outcomes and patient care costs. *Outcomes Manage* 2001; 5(1):36–40.

[3] Arvantes J., *Emergency Room Visits Climb Amid Primary Care Shortages, Study Results Show*, AAFP News, 8/27/2008, http://www.aafp.org/online/en/home/publications/news/news=now/health-of-the-public/2008

[4] Raghupathi W., Data Mining in Health Care. In: Kudyba S., editor. *Healthcare Informatics: Improving Efficiency and Productivity*. 2010. pp. 211–223

[5] Frost & Sullivan: Drowning in Big Data? Reducing Information Technology Complexities and Costs for Healthcare Organizations.

[6] Morgan S., Medication error statistics. *The Prescription* July 2005; 1(1).

[7] Cardinal Health: Statistics – Medication Safety & Education

[8] Joint Commission Patient Safety Goals, http://www.jointcommission.org/PatientSafety/NationalPatientSafetyGoals/09_hap_npsgs.htm

[9] Remko v. d. T., Erik J. v. L., Reinout H., et al. Electromagnetic Interference from radio frequency in critical care medical equipment identification inducing potentially hazardous incidents, *JAMA*. 2008; 299(24):2884–2890 (doi: 10.1001/jama.299.24.2884)

[10] Seidman S. J., Bekdash O., Guag J., Mehryar M., Booth P., Frisch P. H., *Feasibility* results of an electromagnetic compatibility test protocol to evaluate medical devices to radio frequency identification exposure, Biomedical Engineering Open Line 2014, 13:110, online.com/content/13/1/11

Chapter 9

The Role of Real-Time Location Systems in Ambulatory Care

Joanna Wyganowski and Mary Jagim

As communities begin to re-open after COVID pandemic, a "new-normal" began for ambulatory care practices. Ambulatory care practices undertake the difficult task of restoring patient volumes through the practice while limiting exposure risks for patients and staff. This chapter explains the role of real-time location systems (RTLS) in implementing new processes necessary to optimize patient flow while creating a safer environment.

Background and Challenges

Around the globe, the COVID-19 pandemic has caused the illness of millions and a death toll in the hundreds of thousands [1]. The event has altered daily life and impacted businesses. Ambulatory care medical practices are no exception. The operational and financial impact on medical practices has been significant. Many of these practices are smaller and privately owned by providers or groups of providers with fewer financial resources to sustain the practice.

 DOI: 10.4324/9781032690315-9

Impact on Patient Volumes and Revenues

In an effort to reduce potential exposures, many medical practices temporarily closed or had to limit their number of daily appointments, or they transitioned their model of care primarily to telehealth as advised by the Center for Disease Control (CDC) [2]. In some cases, these services have incorporated in-home diagnostic testing such as lab draws or electrocardiograms [3]. These changes significantly reduced their daily patient volumes, compared to pre-COVID-19 volumes. This was highlighted in a survey done by the Medical Group Management Association, in which it was found that medical practices have experienced a 60% average decrease in patient volume and a 55% average decrease in revenue since the beginning of the coronavirus pandemic [4,5].

Impact on Workforce

In addition, ambulatory care practices reported that 48% of them have temporarily furloughed staff, and 22% have permanently laid off staff due to decreased patient volumes and revenues [4].

In those organizations where staff have been furloughed, the staff may choose other opportunities, leaving vacant positions that result in staff shortages once active patient scheduling resumes.

Impact on Workflow and Operations

Ambulatory practices that have remained open and were actively seeing patients have had to alter their existing patient workflows and staff processes in order to maintain social distancing to prevent the potential spread of the virus. All staff had to wear appropriate Personal Protection Equipment (PPE), including a surgical face mask, at all times. Patients were also required to wear a face mask. Screening protocols were put in place to identify potentially infectious patients or staff. Cleaning practices within the facility were hypervigilant and all non-disposable equipment used in patient care had to be thoroughly cleaned after every use [6]. Check-in areas had to allow for ample distancing of 6 feet between patients, and the time patients spent in waiting areas had to be limited or even eliminated [7]. Of further concern was that patients with underlying health conditions, which tend to make up the frequent patient visits to ambulatory care sites, are at increased risk for community-acquired infections.

Patients with cancer and autoimmune diseases were also at higher risk due to their immunocompromised state [8]. These concerns forced practices to implement processes to limit the congregation of patients in areas and the number of patient–staff contact points to reduce exposure risk [9].

As communities began to re-open, a "new-normal" started to begin for ambulatory care practices. There were new challenges as they were restoring patient appointment volumes, including the backlog of appointments, while simultaneously implementing new processes necessary to maintain patient and staff safety [10].

Practices are now faced with new priorities as they start to operate in the post-pandemic era. These priorities include:

- Screening processes to identify patients and staff who may be infectious
- Securing PPE resources for staff and patients
- Maintaining social distancing of 6 feet apart
- Eliminating waiting lines for check-in
- Reducing or eliminating patient time spent in waiting rooms
- Minimizing person-to-person contact while providing the patient with an optimal experience
- Efficiently performing thorough contact tracing to identify patients and staff in the practice at risk in the event of any exposures

Solution

Ambulatory medical practices must develop plans and processes to fully reopen to patients while providing a safe space for patients and staff. Many practices turned to technology for solutions. However, with the economic impact on healthcare, funds for solutions are still limited, so the need for affordable technology solutions has never been greater.

Real-Time Tools

Real-time location systems (RTLS) offer solutions to support ambulatory care practices in overcoming many of these new challenges. Specifically, RTLS tools can provide insight into processes aimed at improving staff workflow and patient throughput and reducing the risk of infections. Traditionally, RTLS solutions tend to be very cost-prohibitive and time-consuming to implement. That is no longer the case with the incorporation of new technology that is more affordable and easier to install. The following are examples of ways in which RTLS can support ambulatory care practices in meeting their goals regarding access and safety.

The Need to Eliminate a Physical Waiting Room and Create Virtual Waiting Rooms

Historically, the reduction or possible elimination of time spent in clinic waiting rooms has always been about patient satisfaction [11]. This includes any waiting in lines to register in addition to the general waiting to be called back to the exam room. However, in the post-COVID-19 era, reducing or eliminating waiting is also about patient safety, by reducing infection risk [10]. As part of reducing the risk of exposure, it is critical to prevent patients from congregating in registration and waiting spaces. Making the transition from a physical waiting environment to a virtual waiting room with contactless check-in is now a top priority for healthcare organizations.

Streamlined Door-to-Exam Patient Flow

An example for improving the door-to-exam room patient flow is to leverage RTLS badges upon a patient's arrival to the clinic. The patient can be automatically placed into a queue to register, allowing the patient to be seated and avoid standing in line, or even to wait in their car; HIPAA-compliant notification tools can alert the patient when it's time for them to register, and with two-way texting, patients can notify the clinic of their arrival or if they're running late. Clinical staff are immediately made aware of the patient's readiness for care to efficiently get the patient into the exam room. RTLS tools have supported ambulatory care centers in eliminating check-in lines and reducing patient wait times by up to 50% [12], creating a safer and more satisfying patient experience. In Figure 9.1, as an example, the staff can readily see the critical information related to the patients' locations and wait times through the Workflow^{RT} solution. In addition, optional tools can be added and utilized for patients to self-room, further improving both patient throughput and safety.

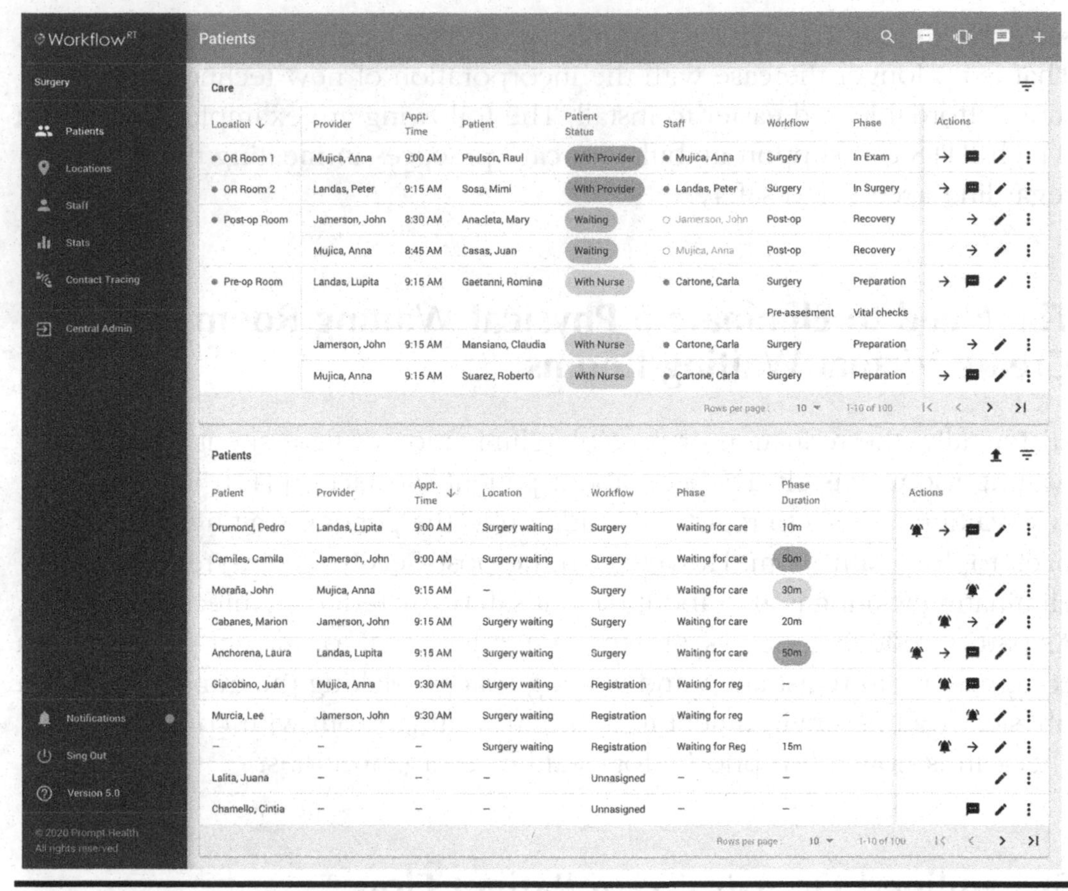

Figure 9.1 Patient flow view in WorkflowRT.

Use of RTLS Data to Improve Staff Workflow and Improve Patient Throughput

As ambulatory care practices begin recovering from the impacts of COVID-19, they need to accommodate as many patients as possible into their schedule, while simultaneously minimizing overall patient time in the ambulatory center. When patients have RTLS badges, staff can quickly know where a patient is, what phase of care they are in, and how long they have been in a specific patient phase (i.e., vital signs, waiting for care, or waiting for registration). Furthermore, with staff wearing RTLS badges, it is also known when a provider/clinician is with a particular patient and for how long. As an example, in the figure below, an onsite patient list from WorkflowRT allows clinicians to see an at-a-glance list of patients waiting to be seen, the order in which they should be seen, and their status. They know which rooms are occupied and which are available. Clinicians are also able to clearly identify how long the patient has

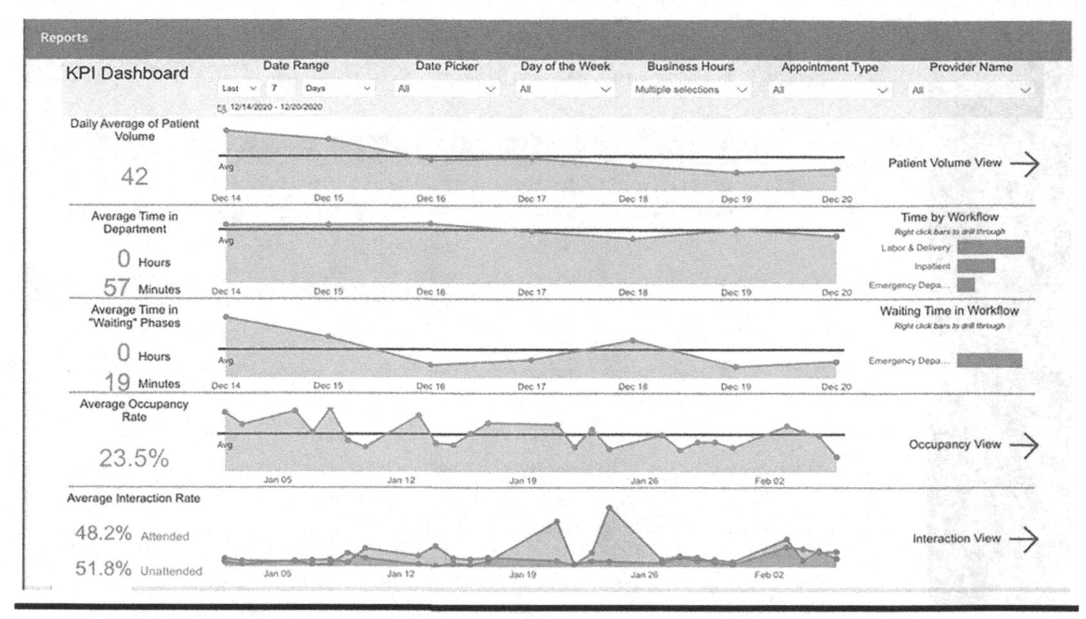

Figure 9.2 Workflow^{RT} KPIs dashboard.

been alone, potentially waiting on a provider or service. Finally, it is also easy to identify which staff the patient is with, and other status indicators. By using RTLS tools, real-time views can highlight the key pieces of information clinicians need to efficiently see patients while minimizing delays. By using these RTLS tools and data effectively to drive staff workflow and patient flow changes, ambulatory clinics have been able to decrease total patient visit time by 8–15%, resulting in a decreased risk for exposures while enabling the opportunity to drive more patient throughput [12].

RTLS can automatically capture data related to location, time, and duration, without requiring manual entry by staff. By leveraging key metrics, ambulatory practices can identify delays in care and modify processes to gain improvements. As a result of workflow efficiencies, ambulatory care practices have been able to increase their patient visit capacity by 5–10% without increasing staff or number of rooms, while at the same time decreasing patient risk of exposure [12]. Figure 9.2 from Workflow^{RT} provides an example of trend data that could be used to improve clinic capacity.

Contact Tracing to Aid in Mitigating Post-COVID-19 Concerns

Part of the "new normal" during the pandemic recovery period, along with future cold and flu seasons, is the ongoing monitoring for COVID-19 cases.

Patient	MRN	Location	Contact Name	Contact Type	Contact Start	Contact End	Duration of Contact
Jenny Smith	1234	Registration Desk	Jeremy Jones	Staff	10:00:00	10:02:00	2 minutes
		Emergency Waiting Room	Jesse Long	Patient	10:03:35	10:05:00	1 minute 35 seconds
			Mary Swanson	Patient	10:04:00	10:14:00	10 minutes
			Dave Gore	Patient	10:04:00	10:08:20	4 minutes 20 seconds
			Julie Woodbury	Patient	10:05:05	10:08:05	3 minutes
		Hallway	Charlies Cannon	Staff	10:07:00	10:07:28	28 seconds
			Beth Briggs	Patient	10:08:00	10:09:12	1 minute 12 seconds
		Triage Room 1	Sue Miller	Staff	10:12:45	10:18:45	6 minutes
		Hallway	Asher Cossette	Patient	10:19:05	10:19:10	5 seconds
		Exam Bay 5	Jayden Staren	Staff	10:25:00	10:32:04	7 minutes 4 seconds
			Pedro Garcia	Staff	10:33:23	10:38:23	5 minutes
		Hallway	John Gibson	Staff	10:36:30	10:38:00	2 minutes 30 seconds
			Paula Perez	Patient	10:40:00	10:41:00	1 minute

Figure 9.3 Contact tracing report in Workflow[RT].

Infection prevention practices continue to be put into place to minimize exposure risk to patients and staff. However, there is also a critical need to be able to conduct efficient, thorough contact tracing for appropriate follow-up when needed. When staff and patients wear RTLS badges, all patient and staff contacts that occurred in the same location, along with the duration of contact time, are monitored. This data enables the ambulatory care practice to promptly take any necessary follow-up action for patients and staff, such as isolation, monitoring for symptoms, and/or prophylactic treatments if appropriate. In Figure 9.3, a sample contact tracing report is shown from Workflow[RT], which quickly summarizes and shows the contacts and contact duration that occurred between patients and staff, along with the location(s). This provides the necessary information for staff to decide on optimal next steps for patient and staff follow-up.

Staying Connected to Families and Visitors

Visitor restrictions have been put in place at many healthcare facilities, meaning patients are either allowed one person to accompany them, or they are required to come alone. Keeping track of visitors, identifying one visitor per patient, and family communication in general can be difficult and time-consuming for staff. The same workflow tools that support patient flow can assist with visitor management and family communication. As an example, the Workflow^{RT} solution allows staff to easily text family members. Each patient can have an unlimited number of recipients, and staff can then send custom messages or select from a list of common messages. This messaging tool keeps families informed while easing the level of effort needed by staff. The same solution can also be used to identify a patient's designated visitor and communicate with them. In addition, with two-way texting, the patient can let the clinic know they've arrived at the hospital, or that they're running late.

Summary

Ambulatory care practices during the pandemic recovery period are experiencing a unique set of challenges. There is a focus on restoring their normal patient volumes while at the same time providing a safe environment for patients and staff that is focused on social distancing and limiting exposure opportunities. Workflow efficiency and heightened awareness of delays are critical when operating in the post-pandemic era.

RTLS-enabled solutions such as Workflow^{RT} provide ambulatory care sites an opportunity to leverage technology to improve their patient experience and decrease wait times while increasing capacity. The workflow solution also helps to reduce exposure points for patients such as those in check-in lines or waiting areas.

The real-time location solution enables ambulatory care sites to:

- Eliminate wait lines for check-in
- Decrease patient wait time
- Allow patients to bypass the waiting room and self-room
- Increase throughput
- Enhance care coordination
- Improve resource utilization
- Provide contact tracing reports for any potential exposures
- Enable family communication
- Provide visitor management

These benefits provide a critical role in ambulatory care in the post-pandemic era, and prepare us to face the challenges that lay ahead.

References

[1] World Health Organization (WHO), (2020). Coronavirus Disease (COVID-19) Outbreak Situation. Retrieved from https://www.who.int/emergencies/diseases/novel-coronavirus-2019

[2] Castellucci, M., Meyer, H. (2020). Physician Practices Modify Operations to Cope with COVID-19. Modern Healthcare, March 17, 2020. Retrieved from Physician practices modify operations to cope with COVID-19 www.modernhealthcare.com›physicians›physician-p ...

[3] Centers for Disease Control (CDC), (2020). Outpatient and Ambulatory Care Settings: Responding to Community Transmission of COVID-19 in the United States. Retrieved from https://www.cdc.gov/coronavirus/2019-ncov/hcp/ambulatory-care-settings.html

[4] Medical Group Management Association (MGMA), (2020). COVID-19 Financial Impact on Medical Practices. Retrieved from https://www.mgma.com/getattachment/9b8be0c2-0744-41bf-864f-04007d6adbd2/2004-G09621D-COVID-Financial-Impact-One-Pager-8-5x11-MW-2.pdf.aspx?lang=en-US&ext=.pdf

[5] Wu Z., McGoogan J. M., (2020). Characteristics of and Important Lessons from the Coronavirus Disease 2019 (COVID-19) Outbreak in China: Summary of a Report of 72 314 Cases from the Chinese Center for Disease Control and Prevention. JAMA. 323(13):1239–1242. doi:10.1001/jama.2020.2648

[6] Centers for Disease Control (CDC), (2020). Interim Infection Prevention and Control Recommendations for Patients with Suspected or Confirmed Coronavirus Disease 2019 (COVID-19) in Healthcare Settings. Retrieved from https://www.cdc.gov/coronavirus/2019-ncov/hcp/infection-control-recommendations.html

[7] Centers for Disease Control (CDC), (2020). Characteristics of Health Care Personnel with COVID-19 — United States, February 12–April 9, 2020. MMWR Morb Mortal Wkly Rep 2020;69:477–481. DOI: 10.15585/mmwr.mm6915e6

[8] Cavallo, J. (2020). Mitigating the Spread of COVID-19 and its Impact on Cancer. *The ASCO Post*. April 10, 2020. Retrieved from https://www.ascopost.com/issues/april-10-2020/mitigating-the-spread-of-covid-19-and-its-impact-on-cancer/

[9] Rivera, A., Ohri, N., Thomas, E., Miller, R., Knoll, M. A. (2020). The Impact of COVID-19 on Radiation Oncology Clinics and Cancer Patients in the U.S. *Advances in Radiation Oncology*, 10.1016/j.adro.2020.03.006. Advance online publication. 10.1016/j.adro.2020.03.006

[10] Centers for Medicare & Medicaid Services. (2020). Recommendations for Re-Opening Facilities to Provide Non-emergency Non-COVID-19 Healthcare: Phase I https://www.cms.gov/files/document/covid-flexibility-reopen-essential-non-covid-services.pdf

[11] Bleustein, C., Rothschild, D. B., Valen, A., Valatis, E., Schweitzer, L., Jones, R. (2014). Wait times, patient satisfaction scores, and the perception of care. *The American Journal of Managed Care*, 20(5):393–400.

[12] Infinite Leap (2020). Ambulatory Care data analytics. Available upon request at http://infiniteleap.net.

Chapter 10

Informatics and Analytics

James Beinlich

The rise of the intelligent health system inherently conjures visions of rich data, deep analysis, and instantaneous on-the-fly decision support or even artificial intelligence. The reality is that many health systems still struggle with multiple core information systems, lack of standardization, lack of governance, and challenges with agreed-upon sources of truth. There is no shortage of data generated by healthcare organizations but making that data actionable at the point of care is still much more of a desire than it is a reality. A recent study by the Virtusa Corporation indicates that healthcare firms typically lag about a decade behind other industries in adopting business technologies that would help with customer engagement. This is not just a technology problem, there are significant people and process issues that contribute to the lag. There is progress and certainly some health systems are seeing success, even integrating patient genetic testing to guide treatment, especially for cancer patients. The recent events associated with COVID-19 have accelerated adoption of technology and data in areas such as telemedicine and use of predictive analytics to determine levels of remote outreach to patients. This chapter will examine the current state of informatics and analytics in healthcare along with the challenges and progress being made.

Legacy Analytics in Healthcare

Until 1983, healthcare reimbursement (in particular, inpatient hospitalization) was a charge-based "fee for service" business. This meant that hospitals billed

DOI: 10.4324/9781032690315-10

insurers (including Medicare) for the fees as a factor of listed charges associated with a hospital stay and were then reimbursed those fees. As medical technology evolved and overall costs for healthcare climbed, the Federal government implemented a new payment system based on DRGs (Diagnosis Related Groupers). DRGs are simply a classification system for patient conditions that can be associated with relative costs and reimbursements. Instead of simply billing the insurer (for non-self-pay patients) for "charges," hospitals now must classify patients into a DRG which has a predetermined payment associated with the DRG. The payment system allows for adjustments based on the severity of patient conditions and other factors (teaching hospitals, indigent care, etc.) but after DRGs, reimbursement was generally no longer tied to charges. Hospitals now had to understand and manage costs as reimbursement had changed dramatically which led to a significant impact on healthcare reporting functions.

There are many things that factor into costs incurred by a hospital or health system including staffing, fixed capital and equipment, supplies (including pharmacy), and administrative overhead. Some of these costs are controllable (staffing and supplies) and thus hospitals began to focus much more intently on understanding how to better manage costs to survive financially. Cost accounting systems were being implemented in every hospital to better understand costs as well as ensure compliance with new federal and state reporting requirements.

Since it is common to see staffing expenses account for about half of the overall costs in a healthcare institution (according to Fitch ratings), it should be no surprise that healthcare began to focus on data analysis related to controlling labor costs. Some staff such as administrative and management are fixed and cannot be flexed but clinical staff (ex. nursing) can be flexed up and down to follow patient volumes and hospital capacity. The better a hospital can flex that variable labor force, the more it can control costs. This activity led to the addition of "management engineers" who were tasked with measuring staff activity and calculating and forecasting ideal staff levels to provide senior management with a level of staffing cost control not needed before the DRG payment system was enacted.

On the supply side of the cost equation, variability in clinical care was identified as a driver of both quality and cost. That meant that those in control of clinical care (mainly physicians) were now being analyzed and judged on their patient care practices to assess and explain variability. Physician scorecards were implemented in many institutions which created a significant amount of friction as physicians were now being made to answer for data reported from the administrative suite as to their clinical practice methods and outcomes. This had a huge impact on the relationship between physicians (many of whom were not employees) and the administrators trying to control the cost driven by those physicians. As you might imagine, the integrity of the data took center stage and was almost always called into question by those being held accountable for what the data was reporting.

For the first time, hospitals embarked on the first real significant endeavor to integrate data from multiple systems for the purpose of conducting true data analysis. Whether it was for clinical quality or financial need purposes, data analysis was rising in importance in healthcare management.

From DRGs to Value-Based Care

Healthcare reform took many shapes and sizes in the years following the implementation of DRGs in 1983. From HMOs to the healthcare reform push during the Clinton administration, costs continued to escalate while questions of quality and outcomes persisted. In 2010, the Affordable Care Act (ACA or Obamacare) once again pushed healthcare to the forefront of national attention. As the ACA expanded coverage for the population, the Federal government via Medicare began to shift risk from insurers to providers. The experience from the HMO era proved that placing risk on insurers clearly did not improve healthcare outcomes for the nation. The term "Accountable Care Organization" or ACO is a value-based payment system that pays providers (and provider/payor organizations) a fee to treat a patient population. The burden of quality outcomes and controlled costs now rests with those deemed best to impact those levers, the providers. Treat a patient for a heart attack, discharge them from the hospital and they are re-admitted 2 days later, the insurer may pay nothing. The reasoning being that the provider must provide a level of care that keeps those situations (costly re-admissions in this case) from occurring.

Value-based Care programs have greatly increased the financial risk to providers, many of whom already operate with very small margins. Given this increased risk, access to accurate, timely, and insightful data on managing these populations and programs has taken on significant importance. It is also important to note that much of the historical scope of analytics for hospitals has been focused on the events that occur when the patient is inside the four walls of the hospital. With outcomes-based programs, however, the focus is on the patient before, during, and after contact with providers. The better managed a chronic or acute condition, the better the outcome and the less "sick-care" a patient will need. Up goes quality, and down goes costs. That's usually how it goes and is the goal of value-based programs – shift from treating sickness to managing wellness.

This change in scope has expanded the realm of data that is required to operate in today's healthcare market. Data related to things such as the social determinants of health (SDoH) now needs to be considered alongside things like lab values, height, weight, and DRG. The CDC considers the following items in its description of the SDoH: life-enhancing resources, such as food supply, housing, economic and social relationships, transportation, education, and healthcare. Combining non-traditional data elements and analysis alongside data points about a patient's health condition (chronic or acute), fundamentally changes the game for data and analytics.

AI in Healthcare: The Holy Grail?

Artificial intelligence (AI) and machine learning (ML) have been touted as a potential game changer since IBM's Watson had its historic Jeopardy performance in 2011. The promise of supercomputers working to solve the puzzle of the world's deadliest diseases (particularly cancer) seemed on the horizon. In 2013, IBM partnered with the MD Anderson Cancer Center, world-renowned for its care and treatment of patients with complex cancers. The goal was to join the oncology know-how of MD Anderson with the power of IBM Watson to identify breakthroughs in diagnoses and treatment of cancer. After five years and $60 million, however, MD Anderson ended the relationship citing "multiple examples of unsafe and incorrect treatment recommendations." Generally viewed as a huge failure by the medical and information technology community, it highlights the need to temper the hype around AI and ML and its true potential in healthcare.

While the failure of IBM Watson in cancer care represents quite a "balloon popping" in the AI world, there are problems in healthcare that AI/ML is showing real promise. As Paddy Padmanabhan from Damo Consulting recently pointed out "AI's performance in healthcare right now is more akin to that of the hedgehog than the fox. The hedgehog can solve one problem at a time, especially when the problem follows familiar patterns discerned in narrow datasets. The success stories in healthcare have been in specific areas such as sepsis and readmissions. Watson's efforts to apply AI in areas such as cancer care may have underestimated the nuances of the challenge."

Given the failure of IBM Watson in cancer care, you might be wondering what are the use cases in healthcare that AI/ML might be successful? There are quite a few, many less sexy and ambitious as helping to cure cancer but much more realistic in delivery potential and thus more practical.

Chatbots

One of the first things that healthcare providers did in response to the early days of COVID-19 was to cancel elective surgeries, procedures, and appointments. The immediate quarantine meant that unless absolutely necessary, face-to-face interactions were a bad idea. Even within the Emergency Room, patients showing symptoms of non-life-threatening were instructed to stay away for fear they could infect others if they were positive for COVID-19. That left a fairly big problem for patients, how to get screening necessary to determine if COVID-19 testing was needed without endangering others by driving to the nearest ER? Enter COVID-19 chatbots. Many organizations have been piloting chatbots for customer service activities and healthcare was no different. What changed was the acceleration from pilot to production.

During the early days of the pandemic, hospital call centers were stretched to their limits contacting patients to cancel appointments and procedures while being inundated with patients and community members calling to get general information about COVID-19. Enter Chatbots. Chatbots were able to pick up the load by answering basic questions about COVID-19 and soon evolved to a more interactive role posing questions and evaluating answers to help determine if a caller had mild, moderate, or severe symptoms. This made it possible to instruct the caller as to the appropriate action to take next (testing, etc.) and provided hospitals with surveillance information as to probable illness severity based on chatbot logs.

Voice Recognition

Voice-recognition systems have come a long way since their introduction in the 1990s. One of the first use cases in healthcare was supplementing physician (particularly radiologists) dictation of reports. Prior to speech recognition software, the process was 100% manual. Physicians dictated their report (usually into a cassette recorder) which was later transcribed by a medical transcriptionist. The transcribed report then had to be reviewed and corrected by the physician before final approval and signature. Voice-recognition software automated much of this process by transcribing the physician's spoken words using pattern recognition and algorithms to guess the words. The physician would then review the output and make necessary corrections. While this improved some aspects of the process (mostly turn-around time and cost savings by reducing transcriptionist labor), early systems were clunky and required many hours of "training" by the speaker to improve the accuracy of the recognition. AI and ML systems have greatly improved since those early days. Companies have consolidated, merged, or been bought by large platform vendors (Google, Microsoft) or new products developed (Amazon Alexa, Siri, etc.). This is greatly enhanced the accuracy and capability compared to early systems and some (Nuance/Microsoft) now tout that their systems can operate in ambient listening mode (aka fly-on-the-wall) and use advanced AI/ML, determine who in the room is speaking (physician or patient) and auto-document the encounter with little/no human intervention.

Imaging

One of the more promising use cases for AI/ML in healthcare is radiomics. As Higgins explains, "When you hear 'radiomics,' think 'advanced imaging analytics.'" AI/ML algorithms can see every tiny particle of a digital image and evaluate change in let's say a lung tumor. AI can determine even the most minute changes in tumor biomarkers related to density, volume, texture, and other characteristics. When compared to prior scans, changes over time become significantly easier to identify. Physicians can also compare these digital markers to thousands of other scans and provide comparative information that can aid physicians in identifying very specific information about tumors and disease progression. Armed with all this additional insight, treatment plans can be tailored for that patient's particular condition and interventions can be enacted much earlier than they would without the insights from AI.

Predictive Analytics

Predictive analytics leverages the patient electronic medical records data (vitals, test results, medications, etc.) and other data (SDoH) along with ML models to monitor (in near real time), patient conditions. The models can then alert caregivers in cases where patient conditions may be deteriorating to assist clinicians in decision making for action to be taken to intervene. Predictive modeling is often done outside of the EMR environment by streaming a live feed of EMR data into the models (running in cloud containers or in on-premises infrastructure), and then directing the output to a separate application for clinicians to monitor or by feeding the output back into the EMR. Feeding the output back into the EMR greatly enhances the usability of the predictive models since that provides the ability to inject decision making directly into workflow in an integrated manner.

Some of the more common models in healthcare include:

- Predicting the onset of Sepsis
- Risk of re-admission
- Risk of inpatient falls
- Risk of Opioid abuse
- Risk of patient no-shows
- Scheduling block optimization
- Heart failure admission risk
- Early detection of stroke

There are many other models that can be developed for healthcare. The list is theoretically endless and only bound by what can be imagined. That is a blessing and a curse, however, as the realities of the "business" of healthcare must also be factored into the equation. There are aspects of those realities that are real barriers to realizing the promise of what AI/ML holds for healthcare.

Barriers to Implementing Predictive Analytics

There are barriers to achieving meaningful implementation of predictive analytics, especially at scale. Running a pilot or simply using it in a single department may prove valuable, but it certainly cannot offer the broad benefit that realizes the true potential of this data and technology. Some of the common barriers are shown in Table 10.1.

Table 10.1 Barriers to Implementing Predictive Analytics

Dimension	*Issue*
Data	■ Lacking enough data to build sufficient models
	■ Bias inherent in ML training models
People	■ Black box syndrome
	■ Data Science staff – cost and availability
	■ Conflicts with existing standards of care
Process	■ Workflow integration challenges
	■ Legal/liability issues (FDA regulation?)
	■ Traditional CDS issues – alert fatigue

Data Issues

Despite having millions of data points from EMR databases, most researchers and data scientists will tell you that there needs to be vastly more data to properly train and maintain ML models. While some large health systems have several million patients in their EMRs, hundreds of millions of patients represent quite a difference when it comes to AI and ML models. Since the training for these models is directly correlated to the number of data points used, it stands to reason that each individual hospital or health system model is limited by the number of its own patient records. This limitation of patient records also adds a significant risk of unintentionally injecting bias into the models themselves. Since a relatively limited data set may reflect only a portion of a population, the characteristics of that population could train a model that is not applicable (or is unfairly biased) to a larger population. EMRs are also notorious for having rich clinical data locked away in unstructured notes. Advanced technologies like Natural Language Processing (NLP) systems are required to unlock even basic information that is mostly unusable in that form.

What can be done? One leading EMR vendor is developing predictive analytic models using data from its large customer base, which accounts for up to 150 million patients in the US alone. Other models are being built as part of multi-site studies that combine several large datasets to improve model development and performance. More work needs to be done to make larger datasets widely available while protecting the privacy of patients, a delicate balance. Data also needs to be free from unstructured notes, either by automated abstraction or minimizing unstructured notes to begin with.

People Issues

One of the issues with AI and ML is "black box syndrome." The details behind AI and ML are not widely understood outside of data science circles. It's a relatively new science, not generally included in any standard business or clinical curriculum (unless you are specializing in this area), and thus many clinical and business leaders don't understand or appreciate what makes up these models. This leads to uneasiness around making significant changes to current processes and procedures (while they may be problematic), yet well understood. Asking an executive (business or clinical) to radically change the way they manage acute and/or chronic disease while not really understanding what's happening in the "black box" is a real problem and one that needs to be addressed before widespread, meaningful implementation of predictive analytics can occur.

What can be done? AI and ML will only be successful in healthcare if it is both transparent and understandable by executive and clinicians. Healthcare is a complex business and if those in charge are expected to trust predictive analytics, it must be clear and easy to understand. One thing that would help is Masterclasses or other certificate programs that could be offered to executives and clinicians to help remove some of the mystery surrounding the "black box." The more healthcare leaders understand how something like this really works, the better the chance they will trust and employ this technology. Another gap exists for resources in the "implementation science" role. Resources that understand both data science technology and healthcare operations could fill the current knowledge gap. By focusing on implementation and adoption of AI/ML technology, the barriers that currently exist would be greatly reduced. This will require investment in what should be expected to be scarce and costly resources, but the downstream potential gain should be considered for making it.

Data science staff costs and availability is another people-related issue that can be a barrier to predictive analytics success. Given the limited understanding of AI and ML at the leadership level, trained data science staff need to help guide the predictive analytics program. The average salary for a data scientist as of this writing is slightly over $100,000 and you probably need more than one. With limited deployment of predictive models and the associated benefits (cost and quality enhancement), it becomes very difficult to justify an investment of this magnitude in staff. Availability is also an issue. According to Quanthub, as of 2020, there is a shortage of 250,000 data scientists in the US. As the demand for data continues to grow (not just healthcare), this shortage should be expected to get worse before it gets better.

What can be done? As is usually the case when resources are scarce in a particular job family, more broad marketing of this job family should be considered at the high-school level. Information Technology programs at the undergraduate level should also be reviewed for opportunities to enhance the data analytics and informatics portions of those degree programs. The recommendation referenced previously regarding Masterclass and certificate programs in data analytics and data science should also be leveraged to increase the available resources that understand this technology.

A final people-related barrier is somewhat related to the prior discussion about the uneasiness of executives to support major changes to existing and well-understood processes. There are established standards of care across every spectrum of healthcare. The definition according to the National Institutes of Health (NIH) is simple "Treatment that is accepted by medical experts as a proper treatment for a certain type of disease and that is widely used by healthcare professionals." This is general but those that practice in specialty care (oncology, cardiology, etc.) are held to an even higher standard. The problem arises with the aspect of care that is described as "widely used by healthcare professionals." This can be in direct conflict with care guidance coming from an algorithm in an ML model or at the very least is not widely enough adopted to give confidence that it meets the standard of care in case of a negative outcome. Even for models not directly associated with clinical care (ex. Risk of no-show), there are tough operational questions to answer. If a patient is flagged as high-risk for now-show by a predictive analytics model, then what? Will that patient's appointment slot be double booked? If so, what are the implications if the patient shows? Will the appointment be canceled? Again, implications? Will extra time and effort be made to contact the patient to confirm the appointment? How many times? What is the cost of the added labor or technology (texting, etc.) to accomplish this? Does the benefit outweigh the cost? Are there issues outside of the patient's control (behavioral health, transportation, etc.)? What should/could be done to address those? As you can see, these issues go much deeper that just running a risk model and require much more serious consideration that one might initially think.

Process Issues

Integration into clinician workflow can be a significant barrier to implanting clinical decision support (CDS) resulting from predictive analytics. Unless your predictive analytics environment is already built into your EMR (EPIC's Cognitive Computing is), then you are faced with making it appear that it is integrated or forcing end users to rely on multiple systems. In an environment where clinician burnout is still a very hot topic, any additional steps (multiple systems, alerts that pop-up during workflow, etc.) are likely to be viewed as just one more thing (and a nuisance) that increases clinician workload. This greatly reduces the chances of acceptable adoption. According to a recent article in EHRIntelligence.com, clinicians at Brigham and Women's hospital were getting roughly one alert for every two medication orders, and clinicians were overriding an astounding 98% of the alerts. "One of the big issues is that many of the clinical systems that are in routine use today, alert too frequently," David Bates, MD, chief of the Division of General Internal Medicine at Brigham and Women's Hospital, said in an interview with EHRIntelligence. "When clinicians are overriding that high a proportion of alerts, clinicians get very used to closing the alert, and sometimes they aren't fully processing what the alerts are saying and they tend to stop paying attention to the important alerts." Add to this sobering fact the "black box" syndrome and you have a real problem with adoption. Until the overall alert issue is dealt with successfully in the EMR, predictive analytics alerts should expect poor adoption.

One final process issue pertains to government regulation of AI/ML technology. In January 2021, the U.S. Food and Drug Administration (FDA) issued the "Artificial Intelligence/Machine Learning (AI/ML)-Based Software as a Medical Device (SaMD) Action Plan" from the Center for Devices and Radiological Health's Digital Health Center of Excellence. As medical device manufacturers (particularly imaging vendors) build AI and ML into their devices, the FDA is developing a regulatory framework to allow its use while ensuring safety and effectiveness. This additional oversight will raise the bar on vendors looking to forward the use of AI and ML but if history serves as a guide, will also slow the introduction and advancement of today's models.

Clinical Data Quality

Healthcare data warehouses that contain EHR data present a significant challenge for research uses due to factors relating to the history of EHRs and the state of database technology and the maturing of analytics in general. EHR data begins its lifecycle as it is entered (manually or via electronic interface) into operational data stores that are generally (and historically) used to optimize capture and storage of transactional activity that supports administrative and clinical process activity. EHR systems have their origins rooted in automation that supports administrative processes related to charging, billing, supply ordering, etc. Historically, rich clinical information has existed in departmental, vertical systems with the EHR providing a broader, longitudinal picture of the patient. This has been changing with the advent of fully integrated EHR systems that provide deep departmental module support in addition to the basic EHR functions but that is a relatively new paradigm that only a minority of health systems enjoy. It is also important to point out that until recent changes in the economic healthcare model (i.e., Accountable Care Organizations (ACOs), Value-based Care), many data points were only the capturing organization's immediate scope of care (i.e., a hospital stay, a clinic visit), leaving gaps in the complete picture of a patient. As healthcare providers are becoming more accountable financially for patients outside of the direct scope of care, more non-event data points are being captured.

Data in the EHR is also limited by the data that is entered into the EMR. As Weiskopf et al. pointed out, the completeness of EHR data for secondary uses is lower than one might expect for a variety of reasons that fall into four general categories; documentation, breadth, density, and predictive.[60] In an effort to address completeness issues that affect secondary use, organizations need to balance the increased need for data with the pushback from busy clinicians who already feel overwhelmed by data entry requirements into the EHR. As Kroth et al. pointed out in a survey of 282 clinicians, 86.9% cited excessive data entry requirements as one of the most prevalent concerns about EHR design and use.

From a pure technology perspective, reporting databases have historically been designed to overcome the limitations (complexity/performance) of operational databases which are optimized to support transactional volume processing. Much like the history of EHRs, reporting databases and systems have traditionally supported administrative and operational reporting activities (capacity, billing, etc.) as opposed to population health and clinical needs, which has resulted in functionality gaps to address things to support ACOs and Value-based Care. Data Warehouses up until recently have been massive, relational DB technologies that do a great job of harmonizing vast amounts of data from a variety of operational stores, but they require significant up-front architecture to optimize performance. This approach forced organizations to "best guess what questions will be asked of the warehouse" to design and implement the warehouse. Overtime, organizations have learned that this, while it provides significant value, is also very inflexible (some changes require re-architecting of the warehouse) and results in long timelines to introduce enhancements. As the move to cloud technology, in-memory processing, and flexible data architecture have evolved, the challenges of legacy data warehouses are quickly being addressed. This is critical to address the fast-paced and ever-changing nature of data analytics and bioinformatics.

Missing data is also a phenomenon with EMRs that has plagued research efforts over the years that is now also causing issues with datasets needed to train ML and AI models. A recent study by Haneuse et al. highlights the issues that missing data can cause such as bias or invalid assumptions. Missing data can be a result of several factors such as incorrect or neglected documentation, incorrect or unreported information by the patient, or missing information as a result of the patient receiving care outside of the host organization (and thus it is not recorded in that EMR). Recent efforts to significantly enhance interoperability should help address the last point but others remain challenging.

Real-time accessibility (data streaming) is another requirement that can be challenging depending on the amount of data being streamed and the ability of local analytics teams to manage these datasets. Streaming data (vital signs and other clinical data from biomedical devices, test results, etc.) require complete uptime and robust data handling to meet the real-time (or near real time) use cases for patient condition surveillance monitoring of some AI and ML models. Streaming data may be a challenge for some IT departments to manage if they do not have the infrastructure and the staff knowledgeable in this area. This is where cloud infrastructure has been very helpful. Most IT departments are well versed in managing relation databases and the associated infrastructure (schema, storage, backups, etc.). Streaming data does not conform to those norms and can be very challenging to manage with legacy on-premises technology. Cloud, on the other hand, natively handles much more than just relational databases. Cloud can manage streaming databases just as easily as relational ones. This allows IT departments to use the same infrastructure to manage both, use service tools in the same environment, and manage data with common tools and processes. This leads to cost savings and reduces the expertise needed to manage the technical environment, making it a real production-ready option.

Genomics and Clinical Decision Support

An exciting new use of data and analytics being placed into the clinical workflow is genomic CDS. By now, many of us have heard the term "precision medicine" or "personalized medicine." Much of what this refers to is the ability to identify and deliver targeted therapies to patients based on their unique makeup. In this case, it's their genetic makeup. Genomic research has been moved forward in a big way over the past 5–10 years, particularly in cancer treatment. Based on research that has identified biomarkers and other commonalities of certain cancers, it is becoming more and more possible to identify patients at higher risk for cancers and to develop targeted therapies for those that have already been diagnosed. Much of this research requires massive amounts of data to conduct Genome-Wide Association Studies (GWAS) and Phenome Wide Association Studies (PheWAS). The data is then analyzed using machine learning and other specialized algorithms that require powerful computing capabilities and lots of storage. This kind of data analytics has led to several developments in precision medicine: drug/gene interaction CDS and genomic CDS.

In drug/gene CDS, clinicians are given real-time alerts if they are about to prescribe a medication to a patient who, based on their genetic makeup, may not benefit from that drug or worse may have an adverse reaction. For example, many patients are prescribed Clopidogrel (better known as Plavix) as an anti-platelet treatment after some cardiac procedures. Plavix requires a certain protein (CYP2C19) which is a liver enzyme to properly metabolize in the body to be effective. In some people, the CYP2C19 gene has variants that cause loss of function, meaning that the process needed to properly regulate that important enzyme does not work correctly. When administered Plavix, the drug will not have the desired effect on the patient and thus place them at higher risk for a cardiac event than those without the genetic variant(s). It stands to reason that if the genetic status of cardiac patients is determined prior to the medication therapy, patients with the CYP2C19 variant could be given an alternative therapy to Plavix, and greatly reduce the risk of an adverse event.

Impact of COVID-19 on Data Analytics

Data and analytics really took center stage in 2020 through the heart of the COVID-19 pandemic. For many healthcare organizations, relying on prior management experience was not much of an option when trying to make decisions in response to the pandemic. The need for real-time data and analytics to support the day-to-day decisions and weekly planning took on incredible importance due to the lack of experience with a pandemic of this scope. Traditional trending and planning tools were no help with managing the dynamic nature of COVID-19, which sometimes changed daily.

As things progressed and some fair amount of data related to COVID-19 patients began to take shape, it became possible to develop basic risk models that could be used to screen patients to assess their immediate healthcare needs whether that was self-treatment at home, or more immediate emergent care. This was mentioned earlier in the chapter in the Chatbot section.

As remote care becomes more of a reality for managing patient chronic conditions as well as post-acute services, risk-based models will become an invaluable tool in identifying which patients are good matches for remote care and which ones are not. Remote care will not be appropriate for every patient and can only be successful if it is applied to those who will benefit.

Data just like other technology is a tool. Using the right tool, at the right time can be a remarkable addition to the healthcare toolkit. Just as any skilled practitioner knows, it takes education, experience, confidence, and trust to use any tool properly to get maximum result. Analytics, informatics, AI, and ML hold tremendous promise and potential for advancing healthcare into the next decade. It's likely that we will be most challenged by the people and process aspects, not the technology.

Bibliography

DuBois, Jen. April 10, 2020. The Data Scientist Shortage in 2020. *Quanthub*. https://quanthub.com/data-scientist-shortage-2020/

Editorial staff. December 10, 2013. 10 Statistics on Hospital Labor Costs as a Percentage of Operating Revenue. *Beckers Hospital CFO Report*. https://www.beckershospitalreview.com/finance/10-statistics-on-hospital-labor-costs-as-a-percentage-of-operating-revenue.html

Haneuse, et al. February 26, 2021. Assessing Missing Data Assumptions in EHR-Based Studies: A Complex and Underappreciated Task [Editorial]. *AMA Netw Open*. 2021;4(2):e210184. doi:10.1001/jamanetworkopen.2021.0184

Kroth, et al. August 6, 2019. Association of Electronic Health Record Design and Use Factors With Clinician Stress and Burnout. *JAMA Netw Open*. 2019;2(8):e199609. doi:10.1001/jamanetworkopen.2019.9609

Landie, Heather. March 30, 2018. Study: Healthcare Lags Other Industries in Digital Transformation, Customer Engagement Tech. *Healthcare Innovation*. https://www.hcinnovationgroup.com/population-health-management/news/13030021/study-healthcare-lags-other-industries-in-digital-transformation-customer-engagement-tech

Miliard, Mike. February 21, 2021. IBM sale of Watson Health could enable renewed focus on cloud growth. *HealthcareIT News.* https://www.healthcareitnews.com/news/ibm-sale-watson-health-could-enable-renewed-focus-cloud-growth

Siwicki, Bill. February 22, 2021. Is AI-enabled radiomics the next frontier in oncology? *HealthcareIT News.* https://www.healthcareitnews.com/news/ai-enabled-radiomics-next-frontier-oncology

Chapter 11

Intelligent Healthcare Use of Germicidal Ultraviolet "C" (UVC) Light

Arthur Kreitenberg

Introduction

This chapter is dedicated to the Infection Control/Prevention professionals to whom we all owe an immeasurable debt of gratitude for selflessly protecting patients, their families, and the institutions they serve.

Ultraviolet disinfection is a powerful tool to disinfect healthcare and other environments. This chapter seeks to educate professionals as to how, where, and when to optimally deploy ultraviolet "C" (UVC) to protect patients and integrate this tool into an overall effective infection control policy. Like chemical disinfectants, a basic science knowledge as well as strict adherence to instructions for use can make the life-saving difference between the appearance of disinfection and actual disinfection.

What Is Germicidal UVC Light?

Ultraviolet germicidal irradiation (UVGI) has been known for nearly one century [1]. It was described as a medical tool in the fight against tuberculosis and measles as early as the 1940s. In recent decades, it has found use as an adjunct to disinfection of healthcare surfaces. SARS-CoV-2 brought germicidal

DOI: 10.4324/9781032690315-11

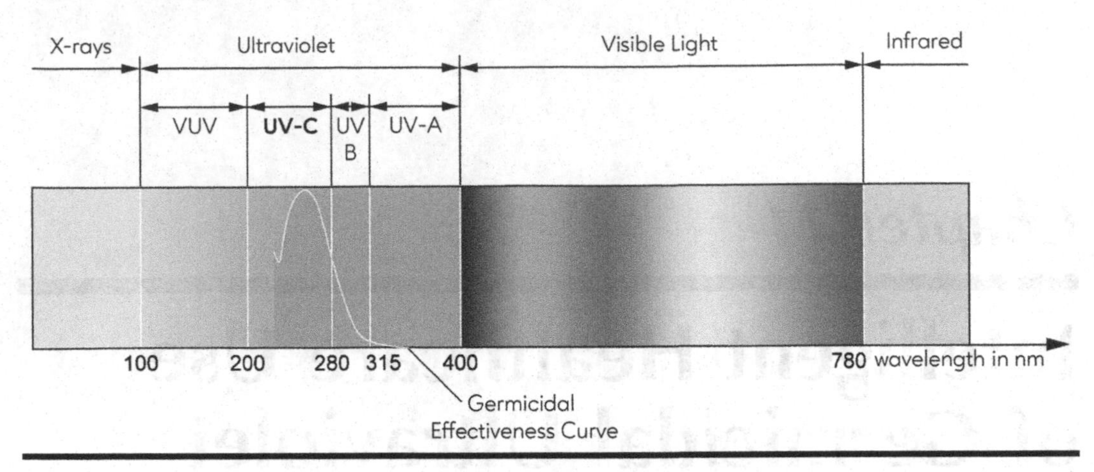

Figure 11.1 Germicidal UV is primarily in the "C" range from 200 to 280 nm.

Courtesy Dimer UV, LLC

UVC to the forefront as a means of disinfection outside of the healthcare environment.

Ultraviolet light lies between visible light and X-ray in the electromagnetic spectrum (Figure 11.1). Ultraviolet light spans a wavelength from about 100 nm to about 400 nm. UVA and UVB, with wavelengths from about 280 nm to about 400 nm, can pass through the Earth's atmosphere. Plant and animal species benefit from UVA and UVB wavelengths of light. UVC, with wavelengths from about 200 nm to 280 nm is readily absorbed by air in the atmosphere and does not reach the Earth's surface. From a teleological viewpoint, living organisms and viruses do not utilize UVC nor do they possess mechanisms to protect themselves from UVC.

UVC has specific adverse effects on biological molecules including nucleic acids that comprise DNA and RNA. When exposed to UVC, the nucleic acids form bonds known as "dimers." Dimers within DNA and RNA preclude normal cell physiology and replication, effectively killing the cell. Because viruses such as SARS-CoV-2 are technically not living, the term deactivation rather than killing is preferred.

Unlike antibiotics and some disinfecting chemicals, there are no known microbes resistant to UVC. UVC is also effective against multidrug-resistant organisms (MDRO) [2]. However, if the UVC is applied at low intensity with slow or incomplete DNA/RNA damage, some bacterial repair mechanisms can overcome the damage caused by UVC, allowing survival to occur. Virus particles do not have repair mechanisms.

Current UVC-Generating Technologies

On earth, UVC is human derived. The peak germicidal effectiveness within the UVC range occurs at about 262 nm[4]. Mercury-based fluorescent lamps are the oldest and most commonly used UVC source and produce peak irradiation at approximately 254 nm. Pulsed xenon emits multiple wavelengths of ultraviolet light within the germicidal UVC range as well as outside the germicidal range. "Far UV" produced with Excimer lamps produces peak wavelengths at approximately 222 nm. UVC generated from light-emitting diodes (LEDs) is a rapidly evolving technology. Very recent advances promise to produce economically viable LED modules in the 265–275 nm range. Each of these sources has advantages and disadvantages that are beyond the scope of this chapter.

Just outside the UVC spectrum are emitters of light at 405 nm. This wavelength disrupts porphyrin, a molecule found within the bacterial cell membrane. Because viruses lack porphyrins, these devices have no specific utility in the fight against viruses, including Norovirus, influenza, and SARS-CoV-2.

UVC Measurement and Specific Pathogen Susceptibility

Units of UVC Measurement

The power of a UVC source may be measured in watts, similar to a household light bulb. A single light bulb is less effective in a large room than in a small room because the same amount of light produced must spread out over a larger volume and surface area. Light is therefore measured in power per area of surface illuminated. For UVC, this is measured in watts per square meter or most commonly in milliwatts per square centimeter, abbreviated as mW/cm^2. This may be considered the intensity of the UVC exposure onto the target surface.

The germicidal UVC dose delivered onto a target surface is determined by the intensity and the number of seconds the target surface is exposed. This is expressed as $mW \cdot sec/cm^2$ or more commonly as millijoules per square centimeter, abbreviated as mJ/cm^2. This may be considered the cumulative or total dose of UVC exposure onto the target surface. For example, a UVC source producing $1 mW/cm^2$ powered on for 10 seconds delivers a cumulative dose of $10 mJ/cm^2$.

UVC Measurement Devices

Dedicated UVC light meters can measure mW/cm^2 output in real time as well as the cumulative mJ/cm^2 for the duration the UVC light remains on

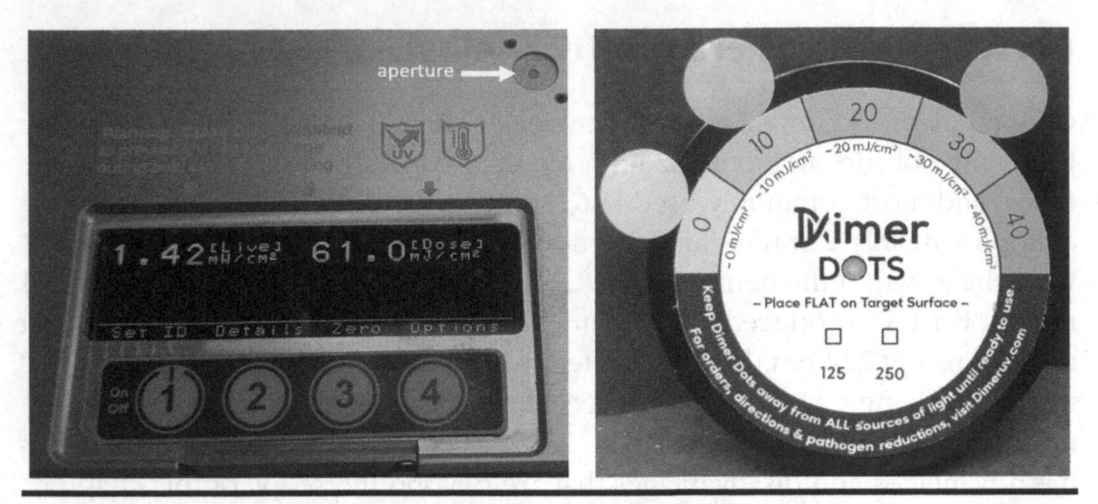

Figure 11.2 UVC cumulative dose is measured in mJ/cm^2 with available dedicated UVC meters. Photochromic indicators are widely used to measure approximate delivered dose.

Courtesy Dimer UV, LLC

(Figure 11.2). These meters tend to be expensive and require regular calibration to maintain accuracy.

One-time use economical photosensitive paper indicators are also commercially available but lack precision and may mislead as they change color with non-germicidal wavelengths of light such as ambient sunlight.

Measurement devices are wavelength specific so a meter that is tuned for a fluorescent lamp at 254 nm may be useless for measuring the dose of Far UVC at 222 nm or an LED UVC source emitting at 270 nm.

To accurately estimate germicidal activity on a horizontal surface, the meter or indicator must be placed horizontally on the surface and not stood up vertically to face the UVC source.

UVC Susceptibilities of Common Healthcare-Associated Infection (HAI) Pathogens

UVC dosing, usually measured in mJ/cm^2, is an approximate guideline of germicidal effectiveness. An excellent compilation of the susceptibilities of multiple microbial species to UVC is available through the International Ultraviolet Association [3]. However, these studies were performed in a variety of environments including air, water, and surface, with differing UVC wavelengths and experimental techniques that may or may not be applicable to real-world situations.

For example, to achieve a 2 log (99%) reduction of Staphylococcus, the required UVC cumulative dose listed in some studies is approximately 6 mJ/cm^2 at 254 nm[5]. However, another study using very high-intensity UVC over less than 1 second achieved a 4.9 log (99.998%) [4] at 6.5 mJ/cm^2 (Figure 11.3). Although cultures remain the gold standard for efficacy, valuable guidance and relative susceptibilities can be gleaned from such data.

Viruses, which can also cause HAIs, are readily deactivated by UVC. Ebola [5], a double-stranded DNA virus, has a susceptibility similar to Staphylococcus. SARS-CoV-2 is a single-stranded RNA virus and appears to be more easily deactivated at lower doses than most microbes. At 254 nm UVC, a 99% reduction of SARS-CoV-2 was achieved [6] at less than 4 mJ/cm^2.

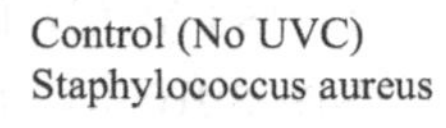

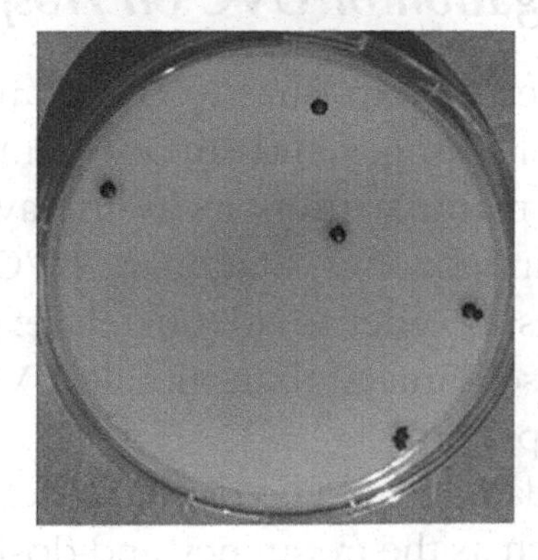

Figure 11.3 UVHammer 21 cm above carriers, control and exposed culture plates.

UVC Safety

Effects and Mitigation of UVC on Humans

UVC, like UVA and UVB, can damage human skin and eyes. The equivalent of sunburn can occur when unprotected skin is exposed to UVC. While there are no known cases of skin cancer caused by prolonged or repeated UVC exposure, deployment of the technology should take this possibility into consideration.

Ultraviolet photokeratitis can occur with relatively low doses of UVC exposure. This is similar to a condition known as "welder's-flash" or "snow-blindness." Affected eyes become symptomatic 12–24 hours after UVC exposure. The discomfort may prompt an ER visit, but the treatment is symptomatic and resolves in 12–24 hours.

"Far UV" at 222 nm, particularly with filters designed to narrow the band of emission, holds great promise as this wavelength appears to be safer for human exposure than other UVC wavelengths. Preliminary studies on mice have been encouraging [7]. As of the time of this writing, Far UV has not been sufficiently studied to determine safe dose thresholds for human skin exposure at 222 nm. The American Conference of Governmental Industrial Hygienists (ACGIH) in 2022 raised the Threshold Limit Value (TLV) to 160 mJ/cm^2 exposure in an 8-hour shift based primarily on animal model studies [6]. Note this threshold is for skin, but not for eyes, so eye protection remains mandatory. Threshold safety numbers are only practical when personal dosimeters are deployed to monitor exposures and ensure safety compliance.

Effects and Mitigation of UVC on Hospital Surfaces

Prolonged and repeated exposure to UVC can degrade some materials and paints causing color changes and surface damage. These effects are effectively mitigated by manufacturers including available UVC stabilizers and special UVC resistant coatings. Stationary UVC devices necessarily overexpose near objects in order to adequately expose distant objects and are more prone to cause damage than mobile UVC devices that provide a more uniform exposure pattern.

To fairly assess damage and color change caused by UVC, it is important to consider factors such as the frequency and dosing of UVC disinfection, as well as the normal aging wear and tear of surfaces in the specific environment. For perspective, effects of UVC on materials must be compared to the effects of repeated application of chemical disinfectants used properly with wet/dwell times and wet/dry cycling.

The Physics of UVC Drives Optimal Device Design

UVC Distance to Target

The distance between the UVC source and the target surface is critical primarily due to the inverse square law (Figure 11.4). Moving the UVC source half the distance to the target surface causes the intensity to increase 4× and the required cumulative dose can be achieved in 1/4 of the time. Similarly, moving the UVC source 3× farther from the target surface will require 9× as long to reach the required dose.

A stationary UVC device will necessarily deliver UVC to objects in a room at varying distances, so nearby objects receive doses far higher than objects further away. Using multiple stationary devices simultaneously or using a stationary device positioned at multiple locations can help mitigate this limitation. A fully mobile device has an infinite number of positions and is most effective at minimizing distances to surfaces in a room.

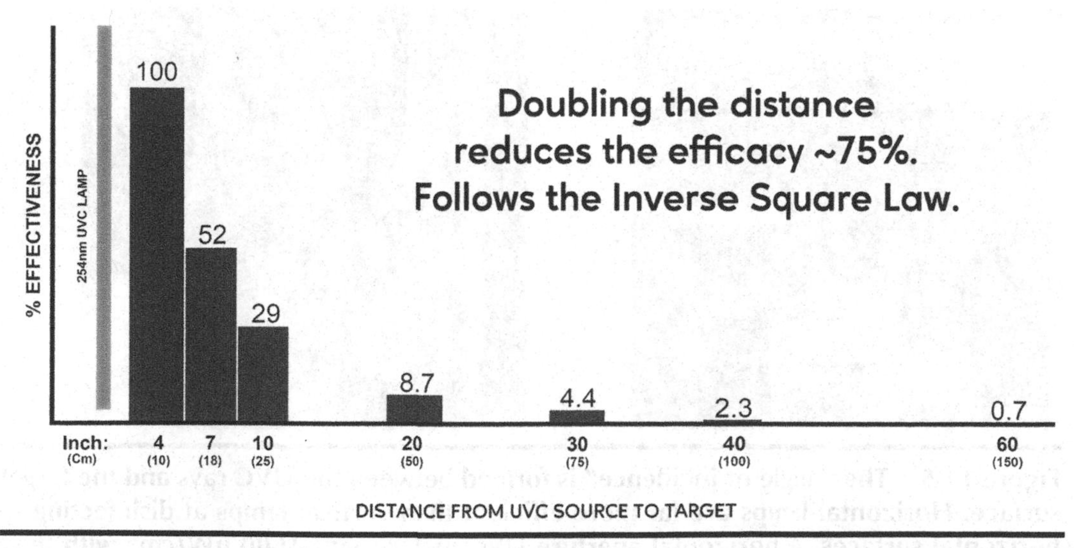

Figure 11.4 The inverse square law dictates light intensity based on the distance between the UVC lamp and the target surface.

Courtesy Dimer UV, LLC

UVC Angle of Incidence

The rays of UVC light emanating from the UVC source can strike the target surface directly perpendicular (90°), tangential to the surface (0°), or at an intermediate angle known as the angle of incidence (Figure 11.5). Stand-up vertical tower UVC devices emit rays predominantly horizontally, parallel to

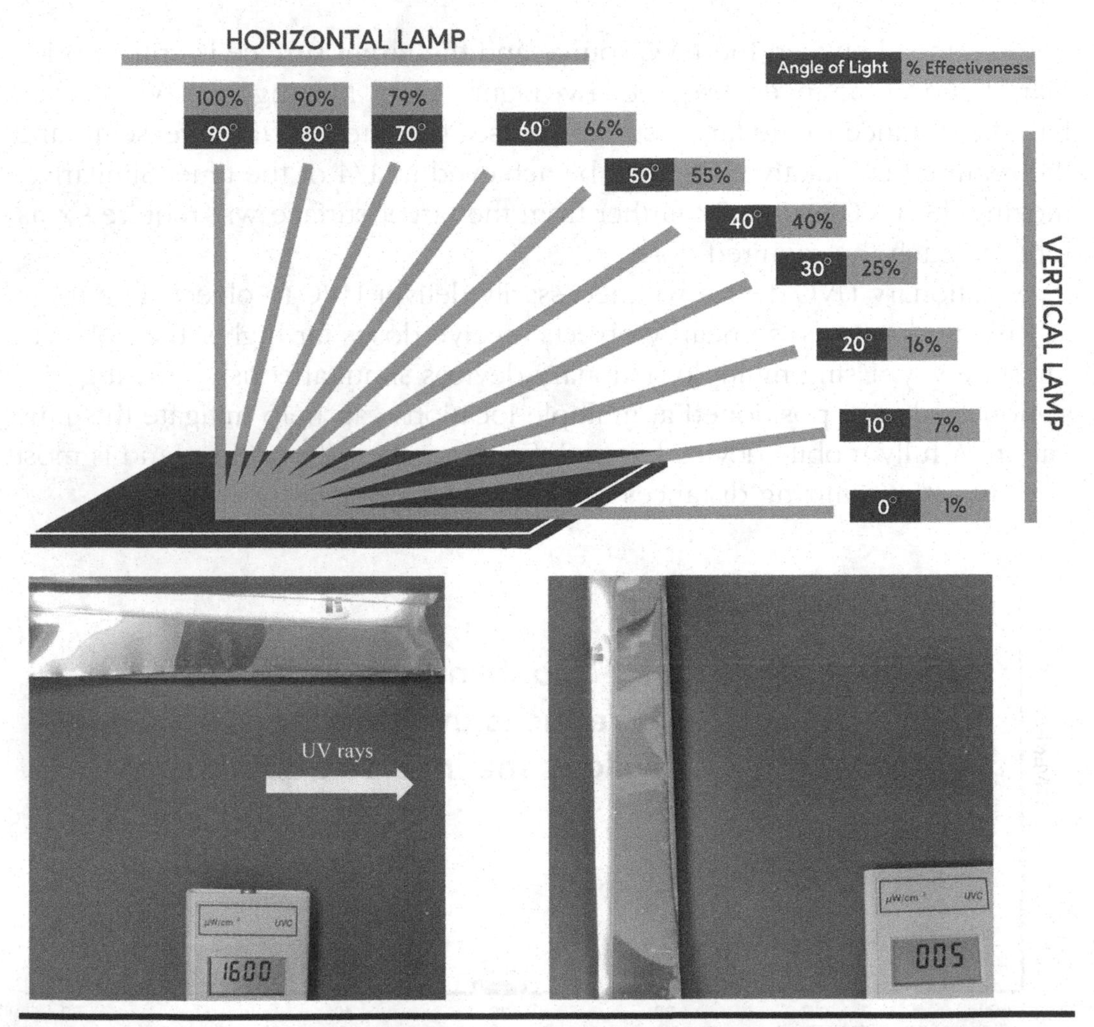

Figure 11.5 The "angle of incidence" is formed between the UVC rays and the target surface. Horizontal lamps are far more efficient than vertical lamps at disinfecting horizontal surfaces. A horizontal aperture UVC meter reads 1600 μW/cm² with the lamp in the horizontal position and only 5 μW/cm² with the same lamp at the same distance in the vertical position.

Courtesy Dimer UV, LLC

the floor and tables. Such devices deliver seven times the dose to vertical surfaces as horizontal surfaces. This results in approximately 100-fold difference in the germicidal effectiveness between vertical and horizontal surfaces with such devices [6]. Like dust, pathogens are far more likely to land onto horizontal than vertical surfaces, placing stand-up tower units at a distinct disadvantage for whole-room disinfection.

Combining the effects of distance and angle of incidence means that lamps placed vertically 1 m from a horizontal surface will require time intervals approximately 1000 times longer than the same lamps placed horizontally 10 cm from the surface to achieve the same germicidal effect. A UVC device that can optimally vary the height and orientation of its UVC lamps relative to the surface orientation is therefore far more efficient at dose delivery and disinfection.

Line of Sight, Shadows, and Poor Reflection

A significant limitation of UVC disinfection is that it is only effective when applied in a direct line of sight. In areas shadowed from the UVC source, doses drop to nearly 0 with effectively no ability to reduce deposited *Clostridioides difficile* [8].

Unlike visible light's ability to reflect off walls, UVC reflects very poorly, on the order of 1–3%, off a typical interior wall. This has practical implications. For example, if an IV pole is to be effectively disinfected with UVC, one cannot depend upon reflection of UVC off the walls to reach the entire circumference of the pole where it is typically grasped. Rather, the UVC device must be placed in at least 3–4 positions around the pole. Similarly, the underside of the surface of a table or chair, which is commonly touched, must be specifically and directly targeted by the UVC device without reliance on floor reflection.

The Canyon Wall Effect

Jaffe [9] observed and confirmed with microbiological experimental data, the UVC shadow phenomenon of textured surfaces at a microscopic level, termed the "canyon wall effect" (Figure 11.6). Textured surfaces include ubiquitous wood, vinyl, Formica, and stainless steel. These surfaces contain "pits and valleys" with depths of approximately 50–100 μm, similar to a human fingerprint and smaller than the diameter of a human hair. Relative to the size of a bacterial pathogen at 1.0 μm, these pits and valleys are the equivalent of a

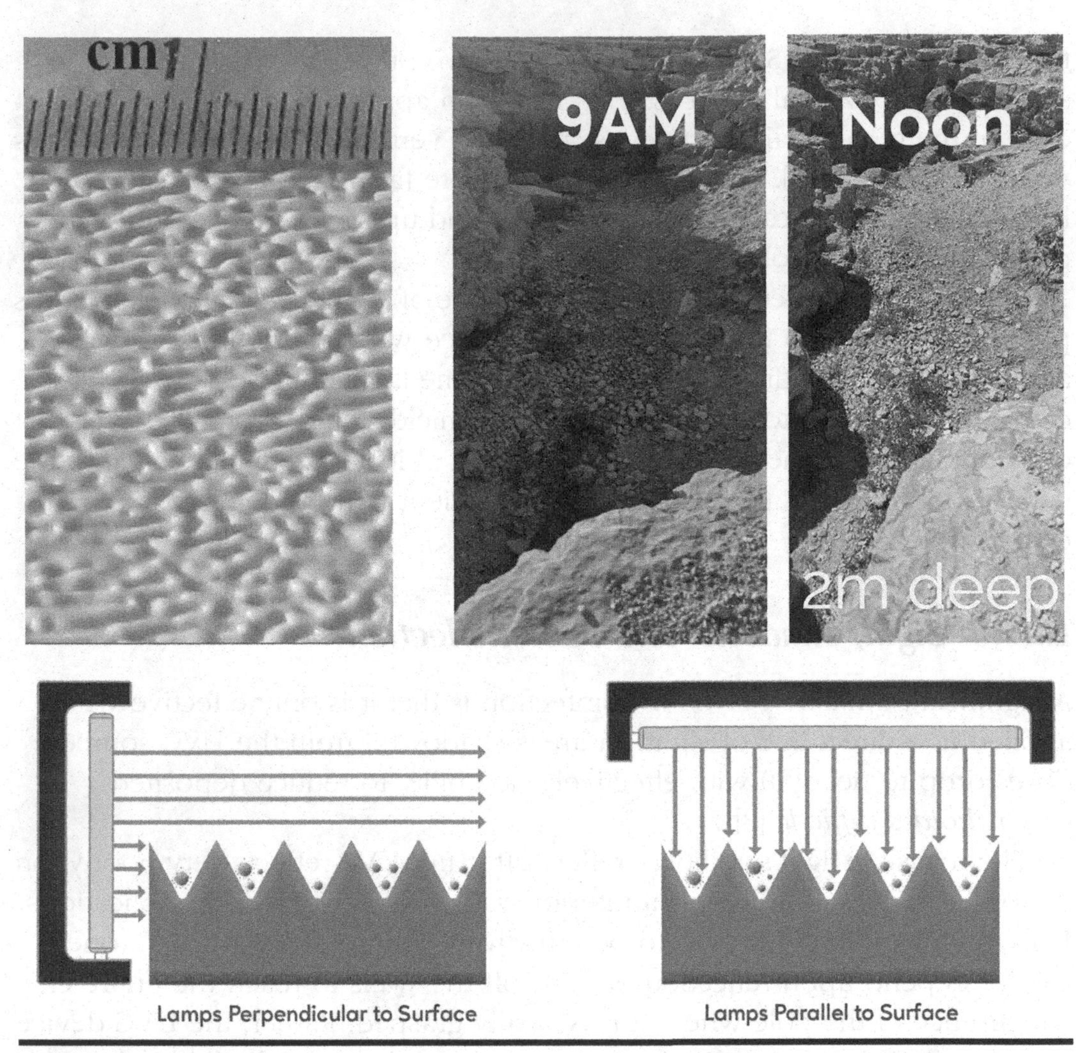

Figure 11.6 The "Canyon Wall Effect" occurs when SARS-CoV-2 particles "hide" deep in a canyon of a textured surface as the light is projected from the side. Smooth Formica has "valleys" about 100 μm deep, with walls 1000x the height of the virus at 0.1 μm.

Courtesy Dimer UV, LLC

human standing in a several hundred-meter-deep canyon. Viruses at 0.1 are the equivalent of a human standing in a several thousand-meter-deep canyon.

Applying a vertically oriented UVC source to a horizontal textured surface is analogous to standing in a canyon at sunrise. The lack of direct line of sight downward into the canyon allows virus particles within the canyon to survive. In contrast, placing the UVC source above the textured surface, like the sun at high noon, allows the germicidal rays to penetrate the canyon floor and deactivate the viral particles.

UVC Transmission and Absorption by Common Materials

Clear or transparent glass and polymers are so-called because they allow transmission of visible light. However, like the Earth's atmosphere, UVC is very effectively blocked by most of these window-type materials. Therefore, a seemingly clear window is like a block wall to UVC rays. Surfaces and air beyond these materials cannot be disinfected with UVC. Similarly, these same materials serve as protective barriers for humans from the potentially harmful effects of UVC which were described above. Some quartz glass transmits both visible and UVC wavelengths.

Current Healthcare UVC Emitter Configurations

There are three current form factors of UVC devices in healthcare: permanently mounted, mobile/positionable, and handheld "wands." Each has a role with strengths and weaknesses. A working knowledge of UVC properties enables the IP professional to logically select from current and future industry offerings.

Ceiling and wall-mounted units emit UVC in a fixed pattern that projects from the device into the room. Target surfaces that face the device can be effectively disinfected, given sufficient time to achieve required dosing. For example, if an overbed table is directly beneath a ceiling-mounted fixture, the top of the table will be adequately dosed with sufficient time. However, frequently touched surfaces such as the side edges and undersurface of the table receive inadequate dosing. The floor around the overbed table will require about double the time to be disinfected, but the floor beneath the table is in the shadows and will remain contaminated. The touched part of an IV pole will not be disinfected, but the air within the beam of the mounted fixture will be disinfected, depending on the distance to the fixture and the time applied.

Mobile/positionable UVC emitters have gained popularity in healthcare with varying rates of proven efficacy. A single-position unit suffers the same limitations as the mounted units noted above. To overcome that limitation, the devices may be placed into multiple locations sequentially, pushed around the room with operator protection, automated on a robotic base, or multiple stationary units may work in conjunction. Such strategies meet with variable success and it is mandatory that the IP professional test the thoroughness of exposure and not rely solely on marketing materials.

Handheld UVC emitting "wands" are available to the general public as well as healthcare professionals. These are useful in tight spaces and complex surface geometries. Many devices are so weak as to be harmless to pathogens and to operators. Those devices that are sufficiently powerful to be effective should be deployed with UVC-blocking PPE.

Limitations of UVC

To optimally deploy UVC, it is critical to understand what it can and cannot do. Based upon the preceding discussion, UVC only works line of sight and on surfaces facing the UVC source. Surfaces not facing the UVC emitter and those blocked by another object will remain contaminated.

UVC will not remove trash, debris, and spills. These require a hands-on robust mechanical solution, usually by trained and dedicated human resources.

Glycocalyx "biofilm" will not be removed by UVC and UVC penetration through an established biofilm is limited [10]. Biofilms require 24–48 hours to form. If UVC is deployed effectively 2–4 times per 24 hours and a daily mechanical chemical wipe is deployed, biofilms formation may be prevented.

Consequences of the Lack of Standards and Regulations

There is an overdue need for regulation of UVC devices and sanctions for false claims to protect the public. The IUVA and ASHRAE (American Society of Heating, Refrigerating, and Air-Conditioning Engineers) are actively working on certification programs for UVC devices.

Even within healthcare, there are neither standards nor regulations for whole-room UVC emitters. The FDA and EPA exclude these devices from their oversight. In 2021, the *Journal of the National Institute of Standards and Technology* (NIST) published physician [11] authored patient-centric recommendations for these devices that can serve as the basis for meaningful standards by the appropriate organizations. Once implemented for healthcare, similar albeit less stringent, standards, and certifications can be modified for community environments at risk.

The lack of standards allows some manufacturers to make claims based on little or no scientific data. For example, a term such as "disinfection cycle time" may be true for vertical surfaces facing the machine. However, such statements are misleading as other surfaces, horizontal or vertical not facing the machine are not disinfected at all.

Although full automation and electronic reporting of cycle time may help "check boxes" for compliance, it remains incumbent upon the Infection Control professional to monitor culture results to determine if a reported "disinfection" cycle or time has actually resulted in disinfection. This need for monitoring will continue until standards are established. Chemical disinfection also requires monitoring for quality control and chemicals do not electronically report to management.

Published studies claiming to lower HAIs should be carefully scrutinized as the root causes of HAIs are multifactorial and complex. Most of these studies are poorly designed and use historical controls and multiple concomitant interventions. There are other studies that show no decrease in HAIs when properly controlled. More reliable studies are those that show decreased bacterial bioburden with the broad inference that a less contaminated environment should result in fewer HAIs.

Standards, regulations, and certifications of UVC emitters can establish the credibility of UVC disinfection as a tool in the prevention and mitigation of HAIs as well as future outbreaks and pandemics.

UVC versus Chemical Disinfection

The EPA regulates chemical disinfectants [12] and defines criteria for "hospital disinfection" with specified log reductions of *Staphylococcus aureus* and *Pseudomonas aeruginosa* but has remained on the sidelines regarding the use of UVC. Emphasis must be placed on following the manufacturer's instructions for use of these chemicals, which are rarely adhered to. This includes proper storage, dilutions, and wet/dwell times that vary from 2 minutes to 10 minutes. It is important that the end-user follow the directions explicitly keeping the surface wet for the required duration which usually involves multiple applications, particularly on vertical or inverted horizontal surfaces. Additionally, a potable water rinse is often required for surfaces that may come into contact with food.

UVC requires no wet/dwell time. Deactivation is instantaneous once the required UVC dose is achieved. Depending upon the UVC equipment deployed, this may take less than one second or more than 10 minutes for a surface to become disinfected.

The EPA approves chemical disinfectants only for "hard nonporous surfaces." Most environments present a variety of surfaces and many are not hard nonporous surfaces. For example only, in a typical hospital patient room, there can be privacy curtains, soft textiles for bedding, upholstered seating, blood pressure cuffs, wood surfaces, and various plastics, metals, and glass. Properly applied UVC excels in disinfecting all such surfaces, potentially supplanting rather than supplementing chemical disinfectant use in some circumstances. When there is a spill, debris, or soiling of an environment, a mechanical wipe with an approved chemical disinfectant, followed by appropriate UVC disinfection may produce optimal results.

Unlike chemicals effective for surface disinfection, UVC is effective for both surface and air disinfection. Whole-room UVC disinfection units, while directing their energy at target surfaces, have the added benefit of disinfecting the air between the UVC lamps and the target surface.

References

[1] Reed NG (2010) The history of ultraviolet germicidal irradiation for air disinfection. *Public Health Rep*, 125(1): 15–27.

[2] Narita K, Asano K, Naito K, Ohashi H, Sasaki M, Morimoto Y, Igarashi T, Nakane A (2020) Ultraviolet C light with wavelength of 222 nm inactivates a wide spectrum of microbial pathogens. *J Hosp Infect*, 105(3): 459–467. 10.1016/j.jhin.2020.03.030

[3] Masjoudi M, Mohseni M, Bolton JR (2021) Sensitivity of bacteria, protozoa, viruses, and other microorganisms to ultraviolet radiation. *J Res Natl Inst Stan*, 126: 126021, available at: 10.6028/jres.126.021.

[4] Kumar A, Pena A, Willner S, Kreitenberg A (21 Dec 2021) UV-C intensity, time and total dose germicidal efficacy, UV solutions. Available at: https://uvsolutionsmag.com/articles/2021/uv-c-intensity-time-and-total-dose-germicidal-efficacy/

[5] Sagrapanti JL, Lytle CD (2011) Sensitivity to ultraviolet radiation of Lassa, vaccinia, and Ebola viruses dried on surfaces. *Arch Virol*, 156: 489–494, DOI 10.1007/s00705-010-0847-1.

[6] Sliney DH, Stuck BE (16 Feb 2021) A need to revise human exposure limits for ultraviolet UV-C radiation. *Photochem Photobiol.* Available at: 10.1111/php.13402.

[7] Buonanno M, Ponnaiya B, Welch D, Stanislauskas M, Randers-Pehrson G, Smilenov L, Lowy FD, Owens DM, Brenner DJ (2017 Apr) Germicidal efficacy and mammalian skin safety of 222-nm UV Light. *Radiat Res*, 187(4): 483–491.

[8] Boyce JM, Donskey CJ (2019) Understanding ultraviolet light surface decontamination in hospital rooms: A primer. *Infect Control Hosp Epidemiol*, 10.1017/ice.2019.161

[9] Jaffe M (2019) UV-C effectiveness and the "canyon wall effect" of textured healthcare environment surfaces. *UV-C Solutions* Quarter, 4: 14–16. Available at https://UV-Csolutionsmag.com/article-archive/digital-archive/.

[10] Argyraki A, Markvart M, Bjorndal LB, Bjarnsholt T Petersen PM (2017) Inactivation of *Pseudomonas aeruginosa* biofilm after ultraviolet light-emitting diode treatment: a comparative study between ultraviolet C and ultraviolet B. *J. Biomed Opt*, 22(6): 065004. Available at: 10.1117/1.JBO.22.6.065004

[11] Kreitenberg A, Martinello RA (2021) Perspectives and recommendations regarding standards for ultraviolet-C whole-room disinfection in healthcare. *J Res Natl Inst Stand Technol*, 126, Article No. 126015 10.6028/jres.126.015.

[12] Product Performance Test Guidelines (February 2018)*OCSPP 810.2200: Disinfectants for Use on Environmental Surfaces. Guidance for Efficacy Testing*. US Environmental Protection Agency, Office of Chemical Safety and Pollution Prevention (7510P) 712-C-17-004.

Chapter 12

Optimizing Infection Control and Hand Hygiene

Matus Knoblich

Introduction

The COVID-19 global pandemic changed people's lives in ways that cannot be overstated. Industries were forever changed, with healthcare policies and procedures being the most directly impacted. As the pandemic swept throughout the world and patients overwhelmed facilities, everyone looked to healthcare to be the savior of lives. With much of the focus on the availability of hospital beds and ventilators, not much of the media focused on what was occurring within healthcare – infection control compliance. Infection control within healthcare effectively started in the late 1840s with Hungarian physician Ignaz Semmelweis, who conducted observational studies and advanced the idea of hand hygiene in medical settings. He proposed the practice of using a chlorinated lime solution to wash hands in Vienna General Hospital's First Obstetrical Clinic, where mortality rates in the midwives' wards fell to three times less than those in the doctors' wards.

Over the years, the practice of hand washing further increased, and in the late twentieth century, healthcare began enacting many of the broader infection control policies we continue to use today. However, the COVID-19 pandemic demanded a re-examination at all levels of infection control to devise new practices and policies to combat the virus in healthcare settings. Despite the mainstream media focusing on hand washing in general, more drastic measures were implemented within healthcare regarding infection

DOI: 10.4324/9781032690315-12

control compliance. This chapter examines the additional infection control measures taken in healthcare settings in response to the COVID-19 pandemic, largely reflected via personal experiences gathered over the past two years as a hands-on biomedical services provider in major healthcare networks.

Hand Hygiene

To understand the overall changes to infection control in healthcare environments during the COVID-19 pandemic, it helps to first review the hand hygiene compliance protocols for infection control implemented by healthcare facilities to combat the COVID-19 pandemic, which put a major focus on hand hygiene. Viruses are easily transmitted, hence hands are a major contributing factor in viral transmission, as patients who touch their faces and other surfaces often can quickly transmit the body from their mouth, nose, or eye area to various surfaces. Though the concept of hand washing in healthcare has been present for over 170 years hand hygiene compliance rates vary. Research has shown that, despite having easy access to hand washing stations and wall-mounted sanitizer dispensers, very few healthcare professionals adhere to proper hand washing and sanitizing practices as dictated by the World Health Organization (WHO)'s "5 moments of hand hygiene":

1. Before touching a patient,
2. Before clean/aseptic procedures,
3. After body fluid exposure/risk,
4. After touching a patient, and
5. After touching patient surroundings.

To combat the lack of hand hygiene compliance , especially during the COVID-19 pandemic, hospitals started by installing additional wall-mounted dispensers. These were installed at the entry of the patient rooms,outside in hallway corridors and in areas that normally would be only presented with visitors. The emphasis was to encourage hand hygiene compliance not only with healthcare staff but also with any visitors to the facility across all areas. Despite the additional installation of these devices throughout facilities, their implementation came with problems. Maintenance of the large volume of devices was problematic, as these devices needed to have the pouches containing sanitizer replaced when empty. Most of these devices do not notify users via any sort of messaging, lighting, or alarming when they are empty; therefore, staff must notify facilities to be changed when they are empty. Unless checked regularly, this creates a deficiency in maintenance. Furthermore, handless versions of these units require regular battery changes to operate properly. When the batteries are not changed in a timely manner, the devices do not operate. The issues of changing sanitizer bladders and batteries were further compounded by staff deficiencies during COVID-19, and an increase in work from additional devices installed to support patient care during the pandemic. As a result, there is a large gap between the number of items needing servicing and the staff available to service them.

And yet, the largest issue facing these devices may have been simply supplying them with hand sanitizer. The pandemic has resulted in a serious strain on supply chains. The lack of employees coming to work due to illness or fear of the illness has resulted in shorter production output. COVID-19 outbreaks within staff at facilities have resulted in shuttering of operations at sites entirely. This has been an issue across all industries and especially in the healthcare industry, notably during 2020. High demands on hand sanitizer coupled with strained supply chains to leave healthcare facilities with shortfalls in hand sanitizer for wall-mounted and other dispensers.

To counter the shortfalls, facilities looked to local and upstart manufacturers of sanitizer, whose products often did not function the same as standard sanitizers regarding hand feel, texture (often very soapy and slick), and smell. Concerns arose about counterfeit sanitizers without the correct alcohol content. Sanitizer needs to be at least 65% alcohol content for efficacy, with 75% being the standard alcohol content to guarantee the killing of all bacteria and viruses. For example, bottled Purell has 70% alcohol content, and the Orbel personal hand sanitizer has 72%. Counterfeit products were found to contain below 50% alcohol content while claiming to contain 75% alcohol. Concerns were also raised that the alcohol used in some counterfeits was not (safe) ethyl alcohol, but methanol, which is toxic and can be absorbed through the skin. The FDA issued warnings and notices alerting users of issues with these sorts of products.

During the pandemic, high demand and low supply led to unsustainable price inflation. Industrial and independent suppliers raised prices on hand sanitizer to unseen levels, peaking at upward of fifteen times their pre-pandemic pricing as demand greatly outstripped supply. With demand coming from commercial, industrial, and consumer sectors, manufacturing was unable to keep up, and pricing became quite volatile. As the pandemic went on, and supply chains were re-established, production returned and even increased with many new market players. To a degree, pricing and supply stabilized; however, 18 months into the pandemic, supply chains remain stressed and supply somewhat limited.

Masks

In addition to hand hygiene, facilities also put a major focus on wearing masks in healthcare facilities. Staff, patients, and visitors were and still are required to wear masks at all times while onsite, regardless of vaccination status. As the virus is easily transmitted through droplets carried by human breath, wearing a mask acts to prevent droplet transmission from a mask wearer to others. The mask, when worn properly, must cover both the nose and the mouth. When breathing, speaking, sneezing, coughing, or otherwise projecting, the droplets will be captured by the mask and prevent transmission from one person to another. When both parties wear masks, this protection is drastically increased, as masks help prevent both the expulsion and inhalation of those droplets.

Healthcare facilities again encountered major obstacles with masks from the onset of the COVID-19 pandemic. The first was availability. As with hand sanitizer, demand for masks came from every sector. Individuals were advised to wear them in the streets, at work, and even at home under certain circumstances. Again, there was huge demand and limited supply.

This was compounded in healthcare facilities, where masks needed to be frequently changed. Should someone be in contact or the vicinity of a COVID-19 patient, as has become very common in healthcare settings, masks need to be discarded and changed when moving between patient rooms in line with contact isolation requirements; this resulted in additional mask demand. Similar supply chain issues that created supply shortfalls for hand sanitizer also affected masks. For example, a majority of mask production was based in China at the onset of the pandemic. With border closings, production shortages, and limited shipping viability from China, supply from main channel factories was limited and at times impossible.

And again, counterfeit products created issues in mainstream supply channels. The Centers for Disease Control (CDC) initially promoted N-95 masks as those necessary for COVID-19 settings; however, due to shortfalls of these masks, many healthcare facilities resorted to 3-ply surgical masks. Supply shortfalls often resulted in healthcare departments purchasing as many masks as possible, wherever possible. Many startup manufacturers seized the opportunity, manufacturing masks that were labeled as N-95 but were not legally certified as such. Significant mask counterfeiting taking place during the pandemic has exposed users to false infection control safety.

Changing CDC guidelines for mask wearing also created issues with mask compliance. When the CDC recommended moving from one mask to double-masking – wearing one mask over another – the rationale was to reduce droplet transmission by increasing layers that droplets would have to pass through to exit into free air. The more layers present, the fewer droplets that can get in or out. Also, due to the propensity of human beings to touch their face and in turn the masks, whether as a habit or to readjust the mask, having a second mask can protect the primary mask from contamination. The outer mask can be replaced more often, protecting the user from direct mask contamination. Minimum two-mask policies were instituted and followed at healthcare facilities in the hardest-hit pandemic regions.

User error with masks presented another major issue in healthcare. The mask must properly cover the user's face over their nose and mouth to provide the greatest efficacy. Unfortunately, many users wear the masks below their noses or below their chin. Whether due to a comfort issue or just plain forgetfulness, this issue of user error allowed the transmission of droplets and caused virus transmission within healthcare.

Room Cleaning

With the major focus being on infection control of the virus via hand hygiene and mask wearing, an area not much talked about outside of healthcare was the implementation of new and updated processes and procedures for room and hallway sanitizing. The virus is easily transmittable through droplets and bodily fluids (sweat, saliva, etc.), which are in turn transmitted through contact with surfaces. Gloves protected the wearer but did not prevent transmission of the virus across surfaces. Therefore, having clean surfaces significantly reduces viral transmission. To combat this, Environmental Services (EVS) departments had to implement additional stringent measures to clean rooms, hallways, and surfaces to keep with changing and tightening infection control measures designed to protect patients, staff, and visitors.

For direct cleaning of surfaces, EVS staff used PDI germicidal disposable wipes to clean all surfaces more frequently than before. Where, in the past, such items would only be used on mattresses and equipment after patient use, new policies implemented the use of PDI Purple and PDI Gray wipes by EVS staff on all surfaces at all times. All medical equipment and items in patient rooms were wiped prior to use, in between procedures, after use, prior to servicing, after servicing, and then finally again prior to patient use. Nurse stations were regularly wiped down, and anything touched by anyone was to be wiped down immediately. The early days of the pandemic resulted in EVS policies that were perhaps overzealousness, wiping every possible surface at any possible time to prevent cross-contamination. As the pandemic entered its second year and COVID-19 transmission was better understood, policies were modified to meet refined CDC and hospital guidelines.

EVS staff implemented a more stringent floor cleaning policy. Hallways and rooms were scrubbed, burnished, vacuumed, and waxed with greater frequency, sometimes every few hours when possible, versus once every cycle as before. Electrostatic sprayers were introduced into much higher use. Demand grew well beyond supply for this equipment, which was very limited; production times became unsustainable. Electrostatic sprayers were rushed to production and broke easily, resulting in additional equipment shortfalls.

At the onset of the pandemic, ultraviolet light devices were mandated for use in all patient rooms after a COVID patient had left. Prior to the pandemic, a typical 600-bed facility might utilize these devices for 200 UV room cleans per month; during the pandemic, this surged to 800–900 UV room cleans per month. Overuse of these devices resulted in frequent burnouts of their lamps, which, due to supply chain shortfalls, were often on back order and difficult to attain, resulting in long repair times for equipment.

Air purification became a greater focus during the pandemic, attracting new manufacturers and new products to the healthcare market. Air purification offered both open and closed-loop systems. Closed systems were preferred because they could be installed in COVID-19 patient rooms and run on a continuous closed-loop system to prevent any air from entering other rooms or sites. However, closed systems were almost impossible to obtain during the first year of the pandemic due to high demand and limited supply. Open systems that existed in the hospital relied heavily on filters, which were quickly in short supply. HEPA filters and vents were overall in such short supply that many were overused past their recommended lifespan, resulting in reduced efficacy in removing harmful agents such as COVID-19 particles from the air.

Overall, EVS and facilities departments within healthcare were limited in their capacity to increase infection control due to a series of pandemic-related shortages. Supply chain shortfalls had a heavy negative impact on the ability of these departments to meet recommended guidelines set forth (and often changing) by the CDC, other governing bodies, and their own healthcare facilities. Equipment maintenance was also an issue, as repair companies were unable to have their staff come onsite to service broken equipment. Staff shortages due to illness and fear of illness, along with limited access due to changing facility guidelines, meant that some companies that would normally come onsite to fix infection control compliance equipment were unable to do so. This resulted in additional downtime to critical cleaning equipment in those instances where replacement parts were available.

Staff

Healthcare staff compliance with infection control measures was perhaps the most important aspect to address during the pandemic. As information on the COVID-19 virus was continuously coming to light, CDC and healthcare facility guidelines reacted. Healthcare staff learned, adhered to, and complied with these infection control protocols. Mask wearing – including how many masks to wear – patient room cleaning and additional cleaning procedures were just a few aspects of that compliance.

Staff education on the new policies and guidelines was key to compliance. Regular training sessions, webinars, emails, fliers, and posters were used as methods to continuously keep the staff apprised of the newest guidelines. However, guidelines often changed so frequently that it created confusion and misunderstanding as to what the most current infection control guidelines were. This situation created infection control compliance shortfalls.

Staff error within guidelines was another major shortfall for infection control compliance. Incorrect hand washing and sanitizing, incorrect mask wearing, not changing gloves often enough, and not cleaning rooms properly all resulted in infection control compliance shortfalls.

COVID-19 infection, of course, created one of the largest issues for healthcare staff. Many healthcare staff were infected and missed work. Others opted out of work or left their jobs due to fear of contracting the virus. This created staff shortages and resulted in overworked staff. Overworked and tired staff were more likely to suffer user error in regard to infection control, allowing for greater risk of transmission of the virus.

Conclusion

The COVID-19 pandemic created serious challenges to infection control in healthcare facilities during the pandemic. While a long overdue reassessment revealed insights like the primacy of hand hygiene to combat infection in healthcare environments, real-life supply chain and staff shortfalls created by the virus resulted in ever-greater risks of transmission. Equipment has been and remains in short supply, and PPE items are in low supply and at risk of counterfeit. Staff are overworked, and guidelines for increased infection control are changing frequently. Despite all this, there has been significant innovation in both equipment and PPE to counter this virus, and the human spirit to persevere through this pandemic has not been deterred.

Case Study: Orbel – Improving the Habit of Hand Hygiene

According to the WHO, there are an estimated two million hospital-acquired infections (HAIs) per year globally, affecting 10% of hospital patients. In Europe alone, the European Center for Disease Prevention and Control estimates that on any given day, 80,000 patients have at least one healthcare-associated infection, i.e., one in every 18 patients in a European hospital. Practicing hand hygiene is a simple and effective way to prevent infections, yet studies show that lack of time, poor accessibility, and high workloads result in a 25% compliance shortfall.[1]

It has been well established that when hand washing compliance can be increased, HAI rates can be reduced. The challenge has been finding ways to effectively increase compliance rates. The idea for the Orbel arose out of this necessity for hand hygiene, coming to market just months before the outbreak of COVID-19 in the United States as a frontline solution for infection control compliance.

Orbel is a patented, wearable hand sanitizer unit purpose-built to increase compliance rates. The orb-shaped unit clips onto your person (i.e., pocket or belt). By gently rolling a hand across its rollerball system, the proper amount of a premium 72% ethyl alcohol gel is released. When rubbed on hands, that gel kills 99.9% of germs within 15 seconds.

Orbel reinforces the habit of hand hygiene via intuitive hand sanitization movement and provides immediate access to compliance in a simple, easy, and cost-effective manner. In turn, Orbel saves lives through improved hand hygiene, reducing the spread of bacteria, viruses, and infection. Current customers include hospitals, restaurants, hotels, banks, schools, professional sports leagues, and law enforcement.

"Orbel has actually been in development for quite a while. A significant amount of research, development, clinical trials, logistics, certifications, and marketing goes into developing a product like the Orbel hand sanitizing solution." Hospitals have capitalized on the value of the product because of how easily it helps increase compliance rates, which is a continual problem they face. With the advent of the COVID-19 pandemic, the general public has become increasingly conscious of hand hygiene. Orbel represents an invaluable tool for anyone that comes into contact with the public or potentially contaminated surfaces, disinfecting with a simple swipe of the hand.

Increasing Compliance at Hospitals

For US hospitals and healthcare facilities, just three words, "The Joint Commission," (TJC) can send waves of panic through the halls. Regulatory compliance is always front of mind, but never more so than during a Joint Commission review, a deep audit performed approximately every three years, during which third-party inspections are performed of the normal daily operations of a hospital unit. Instead of focusing just on paperwork, TJC inspectors scrutinize the delivery of care, treatments, and other services provided by staff, for example, examining Infection Control, Competence, and Environment of Care. TJC's goal is to improve standards across the healthcare industry, including strategies to improve safety and quality of patient care, which is one reason hospitals elect to participate in their reviews.

The most challenging part of the process, perhaps, is that hospitals are provided with little advance warning that an inspection is coming. At most, they are given a seven-day notice, though sometimes it is a last-minute phone call from the lobby, so facilities find it imperative to have their affairs in order to avoid making last-minute improvements.

Orbel is designed to save hospitals and medical facilities significant effort in providing a hand hygiene solution to their staff, doctors, and visitors. The WHO's "five movements of hand hygiene" are widely acknowledged as the best method for preventing healthcare-associated infections. Yet in the critical situations you often find in hospitals, every second counts, and hand hygiene can be de-prioritized. "That's what's most compelling about the Orbel –it provides a point-of-care hand hygiene tool for hospital staff that's literally at their fingertips when they need it. In hospitals, it's important to be able to focus attention on patients, not how they're going to sanitize their hands," states Knoblich. "Hospitals today remain our core customer thanks to the perfect union of efficiency and effectiveness that Orbel provides."

Restaurants, Retail, and Hospitality

The hospitality, retail, and food and beverage industries have been hit hard by the COVID-19 pandemic, subject to increased regulations for cleanliness. These locations face extra scrutiny as crossroads for the transfer of the virus, as a high concentration of people come into close contact with one another, and quite simply, because people often do not take the care to clean their hands enough in these spaces. Bars, restaurants, cafes, hotels, offices, and workshops tend to be ill-prepared when it comes to preventing the spread of infections.

In a hospital setting, healthcare workers are constantly reminded of the need to be clean. It is drummed into their daily practice, yet hand hygiene compliance typically hovers around just 50–60%. Where the issue has yet to be examined, one can suspect even less compliance. Even with a well-trained staff, hand hygiene is not likely to be front of mind. Orbel seeks to address that challenge in any environment where people work together, whether they are serving customers, meeting clients, or just in the same space as other people. In the United States, Orbel intends to meet that need at grocery stores, restaurants, sporting events, and office settings, to provide just a few examples.

Today, infection control is no longer relegated to the realm of hospitals. Rather, it's something that the likes of McDonald's, Yum restaurants, and even Amazon are taking very seriously, presenting dynamic, multifaceted environments where people work together. Whether they are serving customers, meeting clients, or just performing in the same shared space, Orbel represents a solution to personal hand hygiene.

Spreading the Solution: Orbel Distribution Grows

Increased concerns around infection rates continue throughout the world, pushing advances in Orbel from its conception, from reducing packaging to experimenting with different gel formulations. The Middle East, for example, produced unique challenges to ensuring the premium aspect of the Orbel can be maintained globally, i.e., high temperatures that might negatively impact the premium gel solution. Africa, which is set to be the most populous continent on the planet by 2050 and is seeing unprecedented rates of change and growth, presents a different set of challenges to provide a hand hygiene solution in a cost-effective manner and make it as accessible as possible.

Note

1 Gould et al. 2010; IHI 2011

Chapter 13

Hospital at Home: Transformation of an Old Model with Digital Technology

Alisa L. Niksch

The practice of medical care at home is not new. Whether this was a house call for scarlet fever or a centuries-old practice of midwifery, medical care was largely done in the home until the last century. However, the reach of that practice was limited by what a physician or nurse midwife could physically carry in a satchel. For this reason, among other social, technological, and economic advancements, there was a rapid move toward centralization of healthcare. Over a series of decades, however, there were distinct risks which came to be identified within hospital care. Home care began to emerge again, this time for specific clinical purposes, including post-hospital discharge care, palliative care, and rehabilitation services. The span of services has continued to grow, and with the co-development of telemedicine and mobile technologies able to collect and transmit patient data, it has expanded into the care of sicker and more complex patients. Unifying a vetted care model with cutting-edge digital technology and advanced logistics has led to the current state of hospital at home, unifying intelligent patient insights with established cost and outcome benefits to scale this model of care.

DOI: 10.4324/9781032690315-13

Drivers for Change

For decades, the medical industry relied on an increasingly complex epicenter of acute care. However, cost of care for inpatient hospitalizations continues to rise year over year in the United States, rising to $1.1 trillion in 2017 and comprising a third of all healthcare costs [1]. This continued despite the growth of value-based care models and attempts at cost mitigation via service bundling. Meanwhile, clinicians and researchers were working on perfecting a home care model which would provide the same quality of care with the additional benefit of decreasing costs for health systems. Many of these efforts were taking place internationally, in countries like Australia, Canada, Sweden, and Brazil. However, initially, each country may have had different incentives for developing a Hospital at Home model; while Australia appears to have had motivation to offer acute care services in the home to improve system efficiency, certain provinces in Canada may have resorted to acute care at home due to regional bed closures [2].

In the United States, the right technology, care model, and financial incentives are needed to converge to incentivize adoption and expansion of acute care at home by institutions, payers, and patients. The model of care at home in the United States was truly reborn in the early 1970s through efforts of Veterans Administration Medical Centers, with recognition that an aging population of veterans might outpace hospital capacities. The expansion was achieved with further research in academic medical centers, providing data and an opportunity to present this as a feasible option for private payers, and ultimately Medicare.

The Role of the Veterans Administration

In the United States, the Hospital at Home model has been developed the longest within the Veterans Affairs (VA) network, which was able to tie its services to an effective single-payer system for its patient population. The VA system has sustained a robust blend of traditional home nursing care, palliative care, and then developed an acute care at home program prior to its adoption to a broader array of medical centers. In the 1970s, the VA Administration expressed concern that the proportion of aging World War II veterans would expand out of proportion to the general population by the year 2000. These projections led to the development of the VA Hospital-Based Home Care (HBHC) program.

Early evaluations of this program were not complete, they were limited in scope and did not include any cost analysis. Finally, the Edward Hines, Jr. VA Hospital in Illinois, which instituted its HBHC program in 1971, conducted a randomized controlled study on the impact of their physician-led, interdisciplinary home care program on severely disabled or palliative care patients [3]. This study selected patients from those admitted to the 1100-bed hospital, ultimately enrolling 491 to be randomized to usual care vs. home care. The two groups did not have any difference in hospitalization days, though the HBHC group had lower inpatient acuity. The usual care group, however, had much greater utilization of outpatient services. Through mostly lower institutional costs, savings in the HBHC group were 13% lower than the usual care group, with no statistically significant differences in patient outcomes. As would be echoed in future research, patients and caregivers expressed higher levels of satisfaction with the level of care provided in the home, and there was no perceived deterioration in functional status in the patients who received care in a home setting.

This study was one of the hallmark evaluations of a comprehensive care program which involved a multidisciplinary team led by a physician. This was a new model of care which demonstrated longitudinal success and resulted in cost savings to the system. Smaller studies, such as one published in 2018 evaluating a hospital at home program at the VA Hospital in Cincinnati, OH, continued to show benefits in terms of cost and resource utilization, including a lower rate of admissions to nursing homes [4]. This accelerated the adoption of this HBHC model throughout the VA healthcare system and inspired future iterations of the care model and definition of quality metrics. In addition, the early adoption of telemedicine within the VA system in 2003 enabled even more access to care from home and augmented home monitoring for veterans, and by 2014, had reduced inpatient bed days by 59% [5]. This encouraging data, again, served as an incentive to develop future iterations of hospital at home utilizing virtual encounters to enable patient monitoring.

Finding Traction in Academia

The safety and functional impact of hospitalization on the elderly became a focus of Drs. Bruce Leff and John Burton, members of a team of geriatricians at Johns Hopkins Medical Center, in the early 1990s. The team's efforts gained support from the John A. Hartford Foundation, a nonprofit organization which has historically partnered with innovators reexamining care for older adults [6]. Care for older adults in hospital settings has always carried risk of iatrogenic complications including infection, cognitive difficulties, and functional decline. The goal of hospital at home was to prevent these complications, and even enhance the speed of recovery by maintaining mobility and orientation.

Leff's team ultimately developed a program which yielded an early pilot study of hospital at home outside the VA medical system. Published in 1999 in the *Journal of the American Geriatric Society*, the study enrolled 17 patients from the emergency department or outpatient clinic setting. A physician visit was performed daily, and nursing supervision was performed in the home for the first 24 hours, after which a Lifeline Medical Alert system (Philips, Inc., Framingham, MA) was installed. While comparable percentages of patients had services like oxygen therapy, blood cultures, and intravenous antibiotics, "difficult services" at home were noted to be echocardiography, cardiac telemetry, and arterial blood gases. Patient outcomes compared to traditional inpatient hospital care were equivalent for similar diagnoses, and the cost of the acute care at home was 60% of inpatient care [7]. Leff's work led to a keen interest in hospital at home within academic medical centers, even though at the time there was no Medicare coverage available.

Since this time, there have been over 60 publications documenting the clinical benefit and the potential cost savings of acute care performed in the home. A metanalysis of hospital at home data from 61 randomized controlled trials was first published in 2012 in *Medical Journal of Australia*. This looked at metrics regarding mortality, readmission rates, and cost savings among thc multiple studies which qualified. The analysis demonstrated a consistent and statistically significant reduction in mortality and readmission rates. Among the 11 studies which evaluated cost of care, the mean reduction in cost involving 1215 patients was $1567.11 per admission [8].

In November 2020, the Centers for Medicare & Medicaid Services launched the Acute Hospital Care At Home program to provide hospitals a payment structure, and therefore, expanded flexibility to care for patients in their homes. The provisions for the model involved the following requirements:

- Having a physician or advanced practice provider evaluate each patient daily either in-person or through a virtual care platform
- Having a registered nurse (RN) evaluate each patient once daily either in-person or remotely
- Having two in-person visits daily: by either an RN or mobile integrated health paramedics

- Having capacity for immediate, on-demand remote audio connection with a care team member who can immediately connect either an RN or MD to the patient
- Having the ability to respond to a decompensating patient within 30 minutes (e.g., ambulance transfer to an emergency department)
- Tracking defined patient safety metrics with weekly or monthly reporting
- Establishing a local safety committee to review patient safety data
- Ability to provide or contract for other services required during an inpatient hospitalization (e.g., mobile phlebotomy, radiology, or food services).

Documentation of these minimum services is required to fulfill the requirements for Medicare reimbursement. However, the number and variety of services capable of being offered to patients have been the product of innovation, both in the technology and operation sectors. The COVID-19 pandemic certainly accelerated this coverage decision given the massive demand for inpatient care. However, the acceleration of virtual care and adoption of FDA-approved digital medical devices which could substitute for conventional human tasks were the most practical drivers to enable the movement of patients outside a congested medical infrastructure. Meanwhile, the compelling cost and outcomes data had gradually gained traction with private payers. Programs and service providers continue to watch how CMS will view these models as the United States moves beyond the COVID-19 pandemic.

Technology Entering the Home

Virtual care and remote patient monitoring systems had always been an industry of interest to those who were looking for the potential of technology to disrupt healthcare. However, many early attempts at building remote monitoring devices failed to gain traction because of the arduous, expensive, and time-consuming regulatory pathways in the face of unclear consumer adoption prospects. Patients with ongoing health concerns were often left out of the design process, and persuading a medical care team to endorse a non-FDA-approved technology was nearly impossible. Potentially the most impactful piece was the lack of reimbursement structure, either directly or indirectly, which could incentivize the broader incorporation of wearable technology and other data-tracking devices into patient care. It became clear that "disruption" of a highly regulated industry needed a different perspective to succeed.

Since 2017, however, the environment for developing medical-grade remote patient monitoring platforms began to shift. 2017 was the year when the FDA provided official guidance on the handling of medical software and functions of that software which required regulatory oversight [9]. In 2019, the FDA published a subsequent document which gave specific categories of software functions where the agency would not enforce oversight [10]. This maturity of the FDA's understanding of digital medicine helped accelerate the clearance of a large number of newer patient-facing devices, many of them designed for a range of functionalities, from vital sign acquisition to population health applications.

Many of the burgeoning hospital at home programs, as well as contracted service providers, rapidly incorporated these technologies to solve some of their biggest challenges, among them augmentation of care coordination, and reducing the higher cost of human capital to perform lower complexity tasks. Between the devices and the connected platforms which supported them, the technology applied to hospital at home has been key in several ways. For instance, connected devices have become increasingly skillful at informing clinicians of accurate patient data. Telemedicine has increased the number, quality, and efficiency of patient "touches" over the period of acute care performed in the home. Finally, newer health IT infrastructure streamlines clinical workflow, emphasizes technology integration, and creates an intelligent network of coordination of care.

Ian Chiang, an investor at Flare Capital Partners and founding member of CareAllies, which ran a home-based care business within Cigna, spoke about the key features of technology in the home that were particularly valuable in the context of hospital at home:

"From a technology perspective, there are several ways that we need to see continuous improvement. One is a technology platform that can continuously curate diagnostics for the home, especially point of care diagnostics. Second, it's essential to have a well-built API with the layers available to add additional services, and can be deployed to the field in days or weeks. Third, predictive analytics and data science – as the data grows in complexity this will become even more important. Lastly, the user experience for both patient and provider is so important, a technology needs to be functional out of the box to drive engagement [11].

From a product design strategy, the flexibility of the software and API infrastructure is equally, if not more, important than the device it supports. This allows the multifunctionality and versatility of a platform to support monitoring for a range of patient diagnoses and acuity.

Digital Solutions Solving Management Gaps

There is a vast array of continuous vital sign monitors available to support clinical decision within a hospital at home care model. The VitalPatch (VitalConnect, San Jose, CA), which received its 501K clearance in 2018, is a wireless monitor able to detect eight different parameters, including patient position and fall detection. It also provides a continuous single-lead ECG function, which overcomes some of the prior difficulties with continuous cardiac telemetry. Incorporating an access point within the home, and a multipatient continuous monitoring dashboard on the clinician end, VitalConnect clearly markets itself to hospital at home providers.

The all-in-one vital sign wearable from Current Health (Boston, MA, and Edinburgh, UK), extensively used during the pandemic for at-home monitoring, offers an array of vital sign monitoring devices, a tablet for video connection, and a wireless access point. The platform gained FDA approval in 2019 for its platform, as well as its Bluetooth integration with other devices. The company also states that their API has the capability to integrate over 200 additional devices into their platform. They ultimately entered into a partnership with the Mayo Clinic to provide at-home data on convalescing patients with COVID-19 patients who were at risk for deterioration [12]. The company most recently announced in June 2021 that it will become the backbone of the Hospital at Home program at UMass Memorial Hospital (Worcester, MA). These newer companies are competing with more established companies like Royal Philips, which acquired out-of-hospital cardiac monitoring company BioTelemetry in December 2020. Philips had previously contracted with Partners Healthcare to supply their "Hospital to Home" telemedicine and monitoring technology to patients enrolled in various home care programs within the network.

The more complex patient populations are benefiting from rapid development of truly mobile technologies for acute care needs. One of the "difficult" home care tasks noted in early trials was echocardiography, which had no option for portability in the 1990s. However, as the model of healthcare increasingly focused on community outreach, portable equipment became very attractive as an investment. The technology has evolved over the last 20 years from a briefcase-sized device weighing about 6 kg (Vivid IQ, GE Healthcare), to smaller hand-held units weighing about 2–3 kg (VScan, GE Healthcare; Acuson P10, Siemens), to now 0.5 kg probes which can plug into a smartphone (Lumify, Philips; Clarius C3, Clarius Mobile Health) [13]. The Butterfly IQ device is marketed as a hand-held ultrasound device, of which cardiac echocardiography is 1 of 20 presets. The smaller devices tend to have more limited capabilities of color and Doppler imaging, but these new devices have become widely used for point-of-care imaging by specialists outside of cardiology. With the number of options available to clinicians, echocardiography at home is no longer an obstacle to care.

Dialysis has been studied in home care due to a rapidly expanding population of individuals with chronic renal failure. However, most feasibility studies focused on chronic, relatively stable renal failure; a home care nurse would still be required for supervision. One of the more interesting mobile devices in the dialysis space, which was approved in 2019 as a breakthrough device by the FDA, is the AWAK Peritoneal Dialysis Device, which allows the user to self-administer 6–8 hours of treatment through a portable 3 kg device. The device is able to extract peritoneal fluid, filter out toxins through a cartridge integrated into the device, and then infuse the filtered fluid as new dianeal. Traditionally, this process had to be done manually using a new supply of dianeal with each treatment. Other competitors such as Nanodialysis and Triomed are building similar devices which may be on the market in the next couple of years [14].

Overcoming Systemic Challenges

Hospital at home is fundamentally a value-based service, offering acute care comparable to a subset of admitted patients with reduced cost. These types of programs were able to scale in countries with single-payer healthcare systems – Canada and Israel, for example. In the Australian state of Victoria, every urban and regional medical center has a Hospital at Home program, which serves 6% of all admitted patient bed days in the state [15]. In the United States, while this service did have early success under the auspices of the Veterans Administration, it wasn't until the Affordable Care Act incentivized value-based care payment structures that a greater number of private hospital systems grew interested in building programs.

Having a reimbursement structure in place appears to be an essential foundation to the scalability of these services. Hospitals or hospital systems which want to enter into the Acute Hospital Care At Home program had to apply for a waiver on the CMS website. A cohort of six programs were automatically given a waiver after CMS reimbursement was put in place (https://www.cms.gov/files/document/what-are-they-saying-hospital-capacity.pdf); these included Mount Sinai Health System (NY), Massachusetts General Hospital (MA), Brigham Health Home Hospital (MA), Huntsman Cancer Institute (UT), Presbyterian Healthcare Services (NM), and UnityPoint Health (IA). These centers had demonstrated extensive experience with hospital at home services, including publishing results of patient outcomes and cost effectiveness in peer-reviewed journals. However, as of April 2021, there were at least 200 programs which had enrolled in the program. [16]

Many hospitals and hospital systems are still evaluating their capabilities and strategies for taking on increasingly ill patients for home care. Systems like Intermountain Healthcare, based in Utah, have had a foundational home care program since 1984. With experience spanning palliative care to primary care taking place in the home, resources are now in place to take on higher acuity patients. Nickolas Mark, who is a Managing Director and Partner at Intermountain Ventures, sees that "payer reimbursement, provider buy-in, and high-quality coordination of care continue to be the headwinds which may curtail the scalability of hospital-level care at home. However, a demographic of patients over 65, and patient preferences for care after the COVID-19 pandemic are now driving demand for increase home-based medical services, especially in rural catchment areas" [17].

Care Model Execution

Brigham Health officially launched their Hospital at Home program in 2018. Screening patients presenting to their network emergency departments, patients were triaged to acute care at home. The majority of these patients had manageable conditions such as pneumonia, COPD, cellulitis, or urinary tract infections [18]. With an annual volume of about 300 patients, the system was able to demonstrate a significant drop in the need for lab draws and radiologic studies, increased patient mobility and sleep, and an overall drop in cost of 38% compared to traditional inpatient care. Among the 91 patients studied, there was also a significantly lower rate of readmissions (7% vs. 23%) [19].

The history of hospital at home, with years of experience and data showing good outcomes, comparable safety, and cost savings compared to traditional inpatient acute care, justified the investment of larger medical systems in these programs. The programs evolved to partner hospital physicians with contracted services to provide nursing care and monitoring capability, which allowed the care model to scale. Mount Sinai Health System in New York City formed their Visiting Doctor's Program in 1995. This evolved into a Mobile Acute Care Team in 2015, which was funded by a $9.6M Health Care Innovation Award sponsored by CMS [20]. While waiting for a payment plan proposal to be considered by CMS, Mount Sinai ultimately partnered with a third-party service provider, Contessa Health (acquired by Amedisys in 2021), to provide home services and care coordination which were reimbursed by private insurers. Negotiated contracts with several private insurers including their Medicare Advantage plans had been put in place. However, until CMS approved a payment and quality structure for hospital at home in November of 2020, the largest market for these services, those over 65 years old, was often deemed ineligible for this option.

Like Contessa Health, Medically Home is another company providing third-party care coordination which leveraged incipient research supporting a scalable hospital at home model using telemedicine in 2010. It took 5 years until the data was published in the *American Journal of Managed Care* in 2015 [21]. Their data persuaded their first customer, Atrius Health, which was spending hundreds of millions of dollars on hospital care, but did not own a hospital themselves. Medically Home stepped in to provide services, and a revenue stream for Atrius Health. However, it became clear that there were limitations within certain health systems for providing full-service medical care to patients designated appropriate for acute care at home. As the CEO of Medically Home, Rami Karjian, stated:

"While discussing the model with health systems, we quickly realized it wasn't going to develop into thousands of beds across the country by us providing the care. So, we moved to an enablement model. We weren't going to change the country's healthcare if we were going to be the providers" [22].

Karjian also weighed in on the approach to the type of patient who would be an attractive candidate for acute care in the home. He eschewed the idea that home care infrastructure should only be designed for the least acutely ill patients:

> Even though the care is happening in the home, this is hospital-level care in the home, not low acuity home health. You have to build it for high acuity, otherwise you are not going to keep the patient safe, and also you are not going to get the volume … so we said, let's build this for the high acuity patient, and then you can scale all the way down. [23]

There was recognition by the company that in order to gain adoption, the model had to gain trust and assimilate seamlessly into a hospital system. This included utilization of the system's own clinician services and integrating patient-generated data into existing electronic health record software, primarily Epic and Cerner. Adding to the challenge, it was evident that many hospital systems did not have the skills or bandwidth to organize the logistics and supply chain for appropriate care coordination and patient monitoring. Therefore, hospitals could keep their existing payor contracts, maintain a command center with a team of their physicians, but allow Medically Home to deploy technology and personnel to operate an acute care bed at home at a fraction of the cost of care within the walls of a hospital.

The Impact of the COVID-19 Pandemic

The conceptual design of hospital at home models had demonstrated important advantages over traditional inpatient care for many of the common diagnoses now targeted by population health initiatives. However, effective virtual care technologies, which came into the market in the mid-2010s, were essential for the scalability of the model and the ability for clinical staff to access and respond to patient needs at home. This became particularly vital for patient management during the COVID-19 pandemic, where physical human contact had potentially higher risks than benefits.

The COVID-19 pandemic rapidly expanded the patient population which were candidates for hospital at home. However, the pressures brought by waves of infection were not all positive. Some smaller institutions actually contracted or shuttered their home care programs to mobilize clinician resources to intensive care units and other COVID-19 inpatient wards. However, other systems found home care was able to decompress their overloaded inpatient census, conserve personal protective equipment (PPE), and keep infected individuals from mixing with the non-infected on hospital wards [24]. These incentives were also seen internationally, where more or less organized home care efforts were taking place in hard-hit countries like Spain [25] and Italy [26].

Michigan's Metro Health, a health system serving 250,000 patients per year, saw a massive surge in patients infected by the COVID-19 virus in 2020 which threatened to overflow their patient census and safe nursing ratios. When the surge hit their region of the state, Metro Health partnered with Health Recovery Solutions to provide a telehealth and remote patient monitoring infrastructure which was customized for COVID-19 patients. About 20–25% of all COVID-19 patients presenting within Metro Health's system were enrolled in the home care program, and they achieved a 90% adherence to prescribed vital sign measurements [27]. According to the case report published through Health Recovery Solutions, Metro Health prevented an average of 9.5 hospital days for a series of 80 patients.

The COVID-19 pandemic created stress on traditional clinical workflows. In March 2020, CMS made a determination as part of the 1135 waiver that telemedicine visits using audio and video technology would be reimbursed at the same rate as in-person visits, and there were no restrictions to rural geographies. A key change also liberalized the location where patients could receive care, including their own home. No enforcement against virtual visits for initial encounters would take place during the public health emergency, which had not been allowable for reimbursement before 2020 [28].

While telemedicine encounters made up 1.1% of primary care claims for private insurers in Q2 of 2018 and 2019, this skyrocketed to 35.2% of primary care encounters by Q2 of 2020 [29]. A Rock Health consumer survey of 7980 patients in 2020 also documented an 11% increase in patient utilization of live video telemedicine (43% vs. 32%), as well as a 10% increase in the use of wearables (43% vs. 33%) over the prior year. However, the report also revealed that populations least likely to utilize technology to augment or replace conventional models of medical care were in lower socioeconomic groups, residents of rural areas, and those over 55 years of age. More encouraging data was the higher adoption rate within populations with chronic disease states, which are known to drive a significant percentage of healthcare costs in the United States [30]

Another measure that the U.S. government established was a provision in the 2020 CARES Act for a $200M funding program administered by the Federal Communications Commission (FCC). This appropriation, which went into effect on June 25, 2020, was put in place to support and expand telehealth and remote patient monitoring services to improve access to patients during the COVID-19 pandemic. Several centers received funding through this program's first round, including the Mayo Clinic, Ochsner Clinic Foundation, Grady Memorial Hospital, Mt. Sinai Health System (NYC), Hudson River Healthcare, UPMC Children's Hospital, and Neighborhood Health Care (OH). Funding for services, which were invoiced and submitted to the FCC for reimbursement, included the expansion of telehealth and remote patient monitoring services to high-risk homebound patients, focusing on geriatric, palliative, and underinsured patients [31]. A second round of funding to the

FCC for support of telemedicine programs was approved within the Consolidated Appropriations Act of 2021; applications for this second round of funds closed on May 6, 2021.

While outpatient usage of technology to augment access to medical care increased during the COVID-19 pandemic, inpatient care was significantly impacted by the volume of patients inundating regional hospital systems with acute illness. As previously stated, hospital census limitations became a crisis, with many non-COVID-infected patients avoiding or delaying care, or alternatively, competing with COVID-infected patients for hospital beds. This phenomenon likely contributed to the excess deaths tabulated in 2020 [32]. Hospital at Home use grew within certain hospitals and hospital systems but relied on payment systems which fell outside the Medicare reimbursement system. Despite these reimbursement barriers, systems utilizing Hospital at Home like Mount Sinai and Atrium Health tripled their acute home care patient census during the pandemic as a result of excess inpatient numbers [33].

Future Challenges

Hospital at home received a terrific boost from the Medicare coverage determination in 2020. A massive market opened up for patients over 65 years old primarily receiving benefits from Medicare. It appears that the waiver for participation in the Acute Hospital At Home program will be extended until 2022 until further Congressional oversight can take place. However, many participants in the program are systems which already had significant home care services in place and were more easily able to pivot toward a more acutely ill population before and during the peak of the COVID-19 pandemic. Adam Wolfberg, Chief Medical Officer of Current Health, states that even though numbers of institutions participating in the waiver program continue to increase, "larger hospital systems will continue to have an advantage of standing up a Hospital at Home program—they already have the ancillary services set up, or they have existing relationships with contract service providers" [34].

As with telemedicine adoption seen at the beginning of the COVID-19 pandemic, larger medical facilities had a greater proportion of telehealth adoption than smaller practices as a percentage of their pre-pandemic visit volume, reflecting infrastructure and resources which could be shifted to new technology and workflows [35].

Hospital systems that are rapidly moving toward acute care for their home-based care services may also be faced with the challenge of managing remote patient monitoring data. Integrating this data with a workflow that makes sense for the types of patients admitted to hospital at home programs will require time, education, and a strategic focus. This has been a historical dilemma in the commercialization of digital health technologies. The problem of "data dumping" without intelligent curation of actionable information has consistently been an obstacle for timely and accurate clinical evaluation. Again, this speaks to a clinical infrastructure needed to manage data and understand appropriate times to acquire data from patients undergoing home care of acute conditions [36]. The aggregation of even more data from increasingly complex patients will need higher sophistication in data analytics and predictive capabilities to assist clinicians in recognizing potential deterioration in patient status.

In addition, there still exists an educational barrier for patients who may be eligible for this model of care. A level of hesitancy exists regarding the safety and privacy of home care, and building of confidence around this model will be a heavy burden for clinicians recruiting patients for their program. Compounding this hesitancy is an intimidation factor of having to master new technology in the home. While telehealth and remote patient monitoring companies have recognized this barrier and have designed out-of-the-box telehealth kits, some with their own wireless access points, the initial sell to a candidate patient may still be a challenge.

Other temporal factors could also contribute to low patient enrollment into hospital at home programs. A low enrollment rate of 29% was noted at a study site in a 2005 study published in *Annals of Internal Medicine* – this was attributed to a nursing shortage which would have precluded requisite visits at the frequency needed to carry out the protocol [37]. Appropriate personnel with requisite training will take time to develop, even in mature home health programs. The cultivation of a workforce including specially trained paramedics has alleviated the concerns over nursing shortages, and the natural mobility of paramedic duties has proven to adapt to this model of care splendidly. Case managers are also vital to perform intake screening, and they carry a crucial role in identifying candidate patients in outpatient settings, promoting awareness of home care options, and managing rapid deployment of technology and clinical personnel.

Conclusions

Despite the challenges, the demonstration and reproducibility of cost savings to health systems which utilize Hospital at Home is undeniable. This, in addition to the reduction in hospital-acquired infections and functional decline in elderly patients, makes hospital at home an attractive investment, although challenging for smaller institutions with larger infrastructural gaps. Third-party services have started to step in to provide logistical support and coordination of care services, while allowing medical facilities to manage patient care using their own clinical staff. Remote patient monitoring technology supporting the Hospital at Home care model continues to advance in its sophistication. The growing potential of predictive analytics to have an active role in anticipating deterioration in patient status is another driver that can increase the scale of home care for acute conditions. The role of the COVID-19 pandemic has not only shaped patient preferences but also highlighted the value of hospital at home for clinicians who otherwise would have been hesitant to engage with a new workflow. While the reimbursement structure from CMS continues to have an uncertain future, successful implementation of hospital at home services continues to rapidly expand and will create undeniable pressure on agencies to allow these programs to mature.

References

[1] Liang L, Moore B, Soni A National Inpatient Hospital Costs: The Most Expensive Conditions by Payer, 2017. *AHRQ Statistical Brief #261*, July 2020.
[2] Chevreul K, Com-Ruelle L, Midy F, Paris V Issues in Health Economics Newsletter. Institute for Research and Information in Health Economics, Paris, France. December 2004.
[3] Cummings JE, Hughes SL, Weaver FM, et al. Cost-effectiveness of Veterans Administration Hospital-Based Home Care: A Randomized Clinical Trial. *Arch Intern Med.* 1990;150(6):1274–1280.
[4] Cai S, Grubbs A, Makineni R, Kinosian B, Phibbs CS, Intrator O Evaluation of the Cincinnati Veterans Affairs Medical Center Hospital-in-Home Program. *J Am Geriatr Soc.* 2018 Jul;66(7):1392–1398.
[5] "7 Key Findings on VA Telehealth Outcomes", Becker's Hospital Review. June 24, 2014. https://www.beckershospitalreview.com/healthcare-information-technology/7-key-findings-on-va-telehealth-services-outcomes.html
[6] Anthony M Hospital-At-Home. *Home Healthcare Now.* May/June 2021;39(3):127.

[7] Leff B, Burton L, Guido S, Greenough WB, Steinwachs D, Burton JR Home Hospital Program: A Pilot Study. *J Am Geriatr Soc*. 1999 Jun;47(6):697–702.
[8] Caplan GA, Sulaiman NS, Mangin DA, Aimonino Ricauda N, Wilson AD, Barclay L A meta-analysis of "hospital in the home". *Med J Aust*. 2012 Nov 5;197(9):512–519.
[9] "Software as a Medical Device (SAMD): Clinical Evaluation – Guidance for Industry and Food and Drug Administration Staff." Software as a Medical Device Working Group, Food and Drug Administration. Sept 2017. https://www.fda.gov/regulatory-information/search-fda-guidance-documents/software-medical-device-samd-clinical-evaluation
[10] "Policy for Device Software Functions-Guidance for Industry and Food and Drug Administration Staff." Food and Drug Administration. Sept 27, 2019. https://www.fda.gov/regulatory-information/search-fda-guidance-documents/policy-device-software-functions-and-mobile-medical-applications
[11] Chiang I Flare Capital Partners. *Personal interview*, May 24, 2021.
[12] Wholley S Current Health. Mayo Clinic Launch AI-based COVID-19 detection collaboration. *MassDevice*. Apr 29, 2020. https://www.massdevice.com/current-health-mayo-clinic-launch-ai-based-covid-19-detection-collaboration/
[13] Chamsi-Pasha M, et al. Handheld Echocardiography: Current State and Future Perspectives. *Circulation*. Nov 2017;136:2178–2188.
[14] Hu, M Singapore startup's "portable kidney" Can Give Patients Their Freedom Back. *TechInAsia*. Jan 21, 2020. https://www.techinasia.com/singapore-startup-portable-kidney-give-patients-freedom
[15] Montalto M The 500-Bed Hospital That Isn't There: The Victorian Department of Health Review of the Hospital in the Home Program. *Med J Aust*. Nov. 2010;193(10):598–601.
[16] Donlan A CMS Hospital-at-Home Program Closing in on 200 Participants. *Home Healthcare News*. April 19, 2021.
[17] Mark N Managing Director Intermountain Ventures, personal interview, June 18, 2021.
[18] Levine DM, Ouchi K, Blanchfield B *et al.* Hospital-Level Care at Home for Acutely Ill Adults: A Pilot Randomized Controlled Trial. *J Gen Intern Med*. 2018;33:729–736.
[19] Levine DM, Ouchi K, Blanchfield B, Saenz A, Burke K, Paz M, Diamond K, Pu CT, Schnipper JL Hospital-Level Care at Home for Acutely Ill Adults: A Randomized Controlled Trial. *Ann Intern Med*. 2020 Jan 21;172(2): 77–85.
[20] Albert S, Linda DC Inside Mount Sinai's Hospital At Home Program. *Harvard Business Review*. May 10, 2019. https://hbr.org/2019/05/inside-mount-sinais-hospital-at-home-program
[21] Summerfelt WT, et al. Scalable hospital at home with virtual physician visits: pilot study. *Am J Manag Care*. 2015;21(10):675–684.
[22] Rami K CEO, Medically Home. *Personal Interview*, June 2021.
[23] Ibid.

[24] Weiner S Interest in Hospital at Home Explodes During COVID-19. *AAMC News*. Sept 19, 2020. https://www.aamc.org/news-insights/interest-hospital-home-programs-explodes-during-covid-19
[25] Pericàs JM, Cucchiari D, Torrallardona-Murphy O, et al. Hospital at Home for the Management of COVID-19: Preliminary Experience with 63 patients. *Infection*. 2021;49(2):327–332.
[26] Berardi F The Italian Doctor Flattening The Curve by Treating COVID-19 Patients in Their Homes. *Time*. Apr 9, 2020.
[27] Siwicki B Metro Health's Telehealth and RPM Program Is Helping Patients Avoid Hospital Stays. *Healthcare IT News*. June 29, 2021. https://www.healthcareitnews.com/news/metro-healths-telehealth-and-rpm-program-helping-patients-avoid-hospital-stays
[28] "Medicare Telemedicine Health Care Provider Fact Sheet". Centers for Medicare and Medicaid Services Newsroom. May 17, 2020. https://www.cms.gov/newsroom/fact-sheets/medicare-telemedicine-health-care-provider-fact-sheet
[29] Eberly JA, et al. Patient Characteristics Associated with Telemedicine Access for Primary and Specialty Ambulatory Care During the COVID-19 Pandemic. *JAMA Netw Open*. 2020;3(12): e2031640.
[30] DeSilva J, Zweig D Digital Health Consumer Adoption Report 2020. *Rock Health*, April 2021. https://rockhealth.com/reports/digital-health-consumer-adoption-report-2020/
[31] Wicklund E 6 Health Systems Receive Funding From FCC's COVID-19 Telehealth Program. *MHealth Intelligence*. Apr 17, 2020. https://mhealthintelligence.com/news/6-health-systems-receive-funding-from-fccs-covid-19-telehealth-program
[32] Czeisler MÉ, Marynak K, Clarke KEN, et al. Delay or Avoidance of Medical Care Because of COVID-19-Related Concerns – United States, June 2020. *MMWR Morb Mortal Wkly Rep*. 2020;69(36):1250–1257.
[33] Weiner S Interest in hospital-at-home programs explode during COVID-19. *Association of American Medical Colleges*. Sept 29, 2020. https://www.aamc.org/news-insights/interest-hospital-home-programs-explodes-during-covid-19
[34] Wolfberg A Chief Medical Officer, Current Health. *Personal Interview*, June 23, 2021.
[35] Mehrotra A, Wang B, Snyder G Telemedicine: What should the post-Pandemic Regulatory And Payment Landscape Look Like?. *Commonwealth Fund*, August 5, 2020. https://www.commonwealthfund.org/publications/issue-briefs/2020/aug/telemedicine-post-pandemic-regulation
[36] AMA Digital Health Implementation Playbook. American Medical Association Publications, 2018. https://www.ama-assn.org/system/files/2018-12/digital-health-implementation-playbook.pdf
[37] Leff B, Burton L, Mader SL, Naughton B, Burl J, Inouye SK, Greenough WB 3rd, Guido S, Langston C, Frick KD, Steinwachs D, Burton JR Hospital at Home: Feasibility and Outcomes of a Program To Provide Hospital-level Care at Home for Acutely Ill Older Patients. *Ann Intern Med*. 2005 Dec 6;143(11):798–808.

Chapter 14

The Digital OR

Greg Caressi and Bejoy Daniel

Overview

Globally, hospitals and health systems are facing financial pressure due to high operating and administrative costs, declining reimbursement rates, and the increasing cost of both medical devices and infrastructure. A true smart hospital focuses on three major areas: operational efficiency, clinical excellence, and patient centricity, with technological advancements being leveraged to derive smart insights. For clinical excellence, solutions must be implemented so that doctors and nurses can perform their tasks efficiently. Besides using state-of-the-art medical equipment, departments across the hospital need to implement solutions to remove some of the burden from increasingly short-staffed clinical teams, automate processes where possible, and augment human oversight with information and insights, while aligning with best patient outcomes.

Despite the roller coaster ride of COVID-19 challenges, hospitals continue to invest in technology to ensure improved operational efficiency and patient outcomes. A number of foundational shifts occurred, including consumers' increasing involvement in healthcare choices, increasing adoption of virtual health and digital innovations, and the criticality of interoperable data and data analytics. A primary accelerator for these shifts came from the increasing pressure for healthcare providers, governments, payers, and other stakeholders to quickly adapt and innovate to address radical changes in clinical and business conditions during the pandemic.

DOI: 10.4324/9781032690315-14

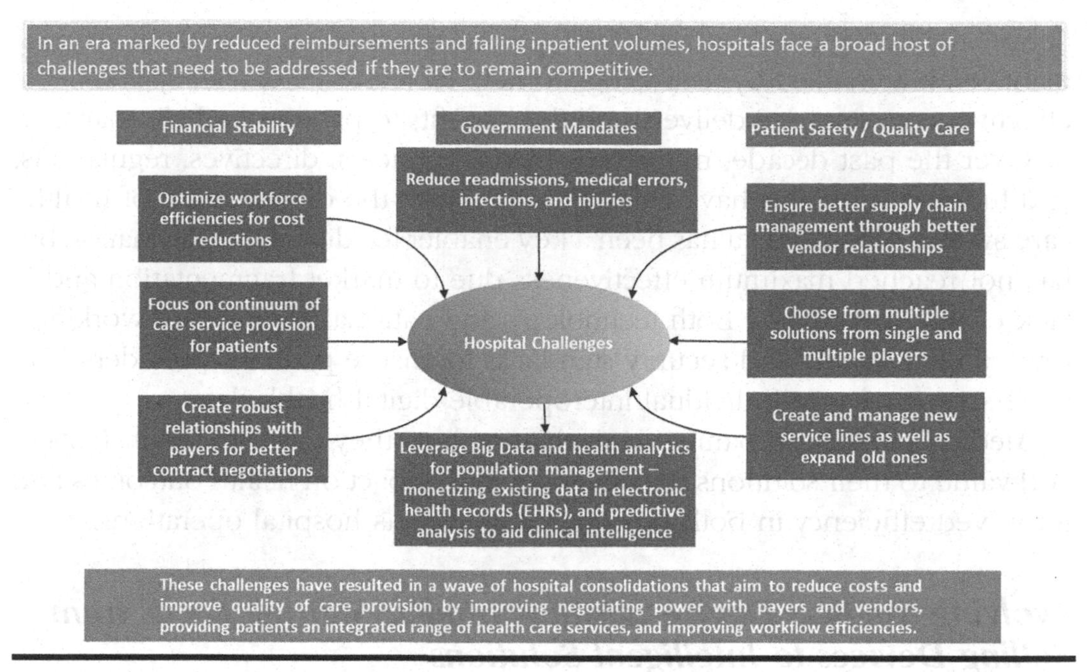
In an era marked by reduced reimbursements and falling inpatient volumes, hospitals face a broad host of challenges that need to be addressed if they are to remain competitive.
Financial Stability
Government Mandates
Patient Safety / Quality Care
Optimize workforce efficiencies for cost reductions
Reduce readmissions, medical errors, infections, and injuries
Ensure better supply chain management through better vendor relationships
Focus on continuum of care service provision for patients
Hospital Challenges
Choose from multiple solutions from single and multiple players
Create robust relationships with payers for better contract negotiations
Leverage Big Data and health analytics for population management – monetizing existing data in electronic health records (EHRs), and predictive analysis to aid clinical intelligence
Create and manage new service lines as well as expand old ones
These challenges have resulted in a wave of hospital consolidations that aim to reduce costs and improve quality of care provision by improving negotiating power with payers and vendors, providing patients an integrated range of health care services, and improving workflow efficiencies.

Figure 14.1 Hospital challenges.

Source: Frost & Sullivan

An additional transformation that is ongoing is the move toward the provision of care outside hospital settings. A larger number of surgeries and diagnostic procedures that historically required an inpatient hospital stay are now being evaluated for safe and efficient performance in an outpatient setting, ranging from ambulatory surgical centers to patients' homes (Figure 14.1).

Digital Transformation Shaping the Future of the Medical Technology Industry

With consumers and patients seeking on-demand healthcare services at their own convenience, the healthcare industry is entering the era of digital innovation. Healthcare systems are facing unprecedented pressure, as many clinicians are struggling to cope with increasing workloads and the gap between the supply of resources and the demand for healthcare is widening. Most countries are looking to digital transformation to close this gap, but progress has been slow, and the digital maturity of providers, both within and between countries, varies widely.

Digital technologies can integrate care, identify and reduce risks, predict and help manage population health needs, and improve the quality of data flow to deliver timely, efficient, and safe care. Digital transformation is about change management enabled by technology to help increase the efficiency and effectiveness of service delivery and the benefits to patients and clinicians.

Over the past decade, numerous national policies, directives, regulations, and funding programs have emerged to support the digitalization of healthcare systems. Health data has been a key enabler for digital transformation but has not reached maximum effectiveness due to market fragmentation and lack of interoperability. Both technology and national policies are working toward data quality and security standards to ensure patients, providers, and payers get access to individual interoperable digital health data.

Medical device companies are evaluating how they can use data insights to add value to their solutions to deliver a wider impact on health outcomes and improved efficiency in both procedures as well as hospital operations.

Evolving Medical Device Business Model: Shifting Focus from Selling Devices to Intelligent Solutions

Until recently, medtech companies succeeded by ensuring that hospitals remain updated with incremental improvements to technology at gradually increasing prices. Competition was driven by delivering the next technology iteration in a similar package.

Medtech companies are now being forced to rethink this "product-centric" strategy. In response to the shifting incentives to healthcare providers and insurers, many medical device companies are evaluating the addition of services or a segmentation strategy for value-based products to broaden their value proposition to clinicians, hospital executives, and administrators.

The medical device industry is struggling to reinvent itself, with increased pressure on medical device and diagnostics product pricing, continued consolidation of buyers, and hospitals seeking to reduce the number of vendors supporting their systems. Hospitals are seeking to improve their operational efficiency through lean principles and predictive analytics.

This is contributing to the trend toward relationships between medtech vendors and hospitals moving from supplier toward partnership in helping health systems meet their healthcare goals. Medtech vendors are being asked to provide their customers with data analytics and predictive capabilities for procurement and inventory management. Vendors that offer a broad product portfolio with patient data outputs useful to hospitals will likely be preferred.

To become streamlined and cost-effective, healthcare providers need to consistently make excellent operational decisions. Hence it becomes critical

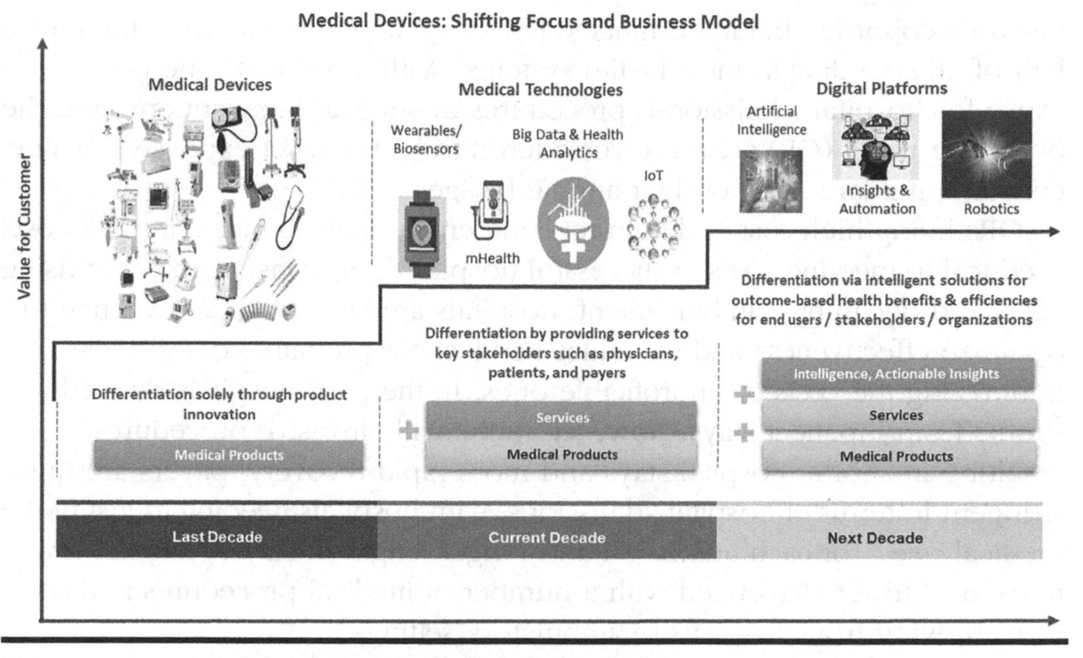

Figure 14.2 Evolving medical device business model.

Source: Frost & Sullivan

for hospitals to establish an effective operational management process through a centralized command-and-control capability. An effective process would ideally be one that is predictive, uses optimization algorithms and artificial intelligence to deliver prescriptive recommendations, and at the same time continues to learn. The ability to mine and process large quantities of data to deliver recommendations will help in streamlining hospital operations.

The shift to intelligent solutions will see improvement in hospital operational efficiency through predictive analytics to improve planning and execution of key care delivery processes. Apart from execution, resource utilization and staff scheduling, and patient admission and discharge will help improve revenue, lower cost, and increase asset utilization to deliver a better patient experience (Figure 14.2).

Challenges in the Operating Room

Providing good quality care has always been the primary objective and priority of healthcare institutions worldwide. With the rapid development of medical technologies and their application, healthcare expenditures have accelerated at rapid rates globally, challenging health systems, private insurers, and governments. National health spending around the world varies between countries, but it is estimated that hospital budgets account for almost half of all spending in most health systems. Within hospitals, the primary cause for hospital admission is procedures or surgical interventions, and the operating room (OR) areas are considered to be the most expensive facility, consuming a large part of their annual budget.

ORs being high-cost and high-revenue environments, managing OR costs is critical to meeting goals of successful hospital operations. In an era of rising costs and declining reimbursement, hospitals are faced with a challenge to optimize effectiveness and maximize services for profitable cases while minimizing the costs of unprofitable ones. In the past, hospitals derived profits from inpatient stays. Now, with minimally invasive procedures resulting in shorter hospital stays and more rapid recovery, payers are quite stringent in terms of hospital admissions, with hospitals needing to justify the medical need for each admission and longer length of stays. Hospital revenues are further challenged with a number of medical procedures and services moving from hospitals to ambulatory settings.

OR: An Important Profit Center

In the average hospital, ORs are estimated to drive approximately 60% of revenue and 40% of total hospital expenses. On average, hospital costs for stays with OR procedures are more than twice the cost for an inpatient stay without an OR procedure. An hour in an OR – depending on the procedure – can cost $150–250 per minute, excluding material costs.

The OR being a profit center, hospitals seek to manage costs and simultaneously increase revenues by ensuring efficient actions by well-trained surgeons and OR staff, effective scheduling, and monitoring of overall OR performance. Some of the most critical hospital issues entail finding new methods to reduce human errors and improve efficiency and quality of surgical care. New technologies are made available to help hospitals reduce risks from surgical procedures while maximizing their return on OR efficiencies. OR efficiency is significant to the field of public health because optimization of efficiency contributes to the hospital's financial success and long-term viability in serving the community's needs.

Reaching these goals on a sustainable basis in the modern healthcare environment requires digital transformation, with challenges similar to what we see with digital health – new financial investments, the absolute need for

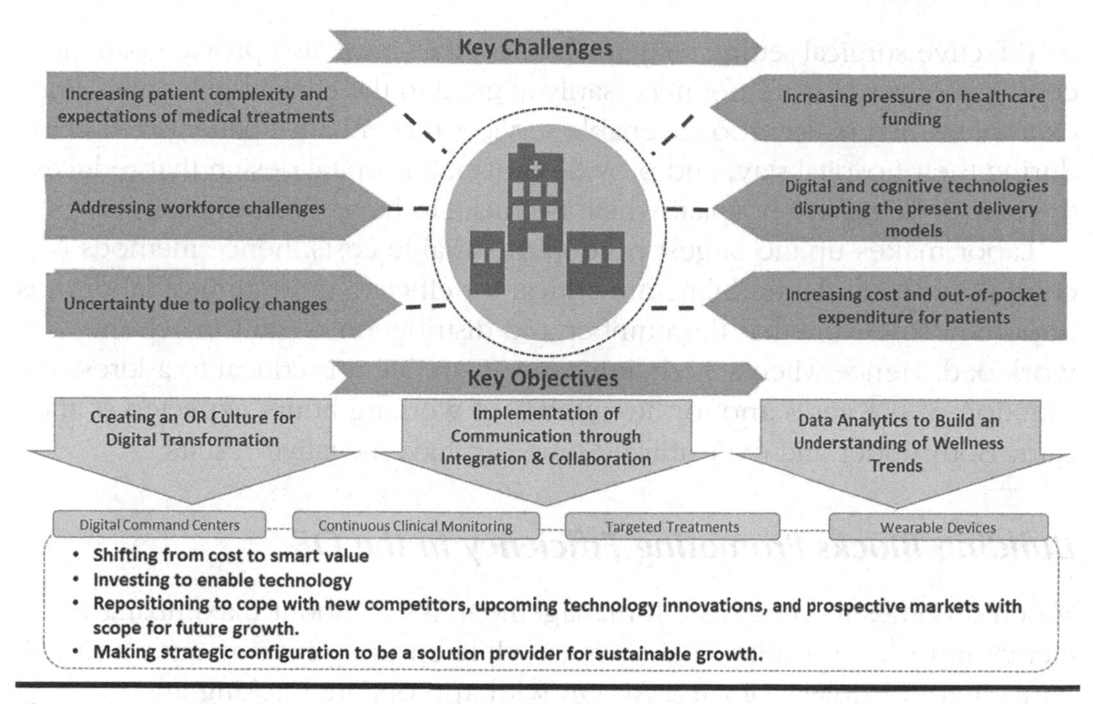

Figure 14.3 Key objectives for stakeholders for ORs.

Source: Frost & Sullivan

interoperability, cybersecurity risks, and a lack of expert resources. Hospitals currently struggle with the major question of where to begin, hence a detailed, comprehensive solution set needs to be formulated for the hospital as a strategy plan, keeping systems open and interoperable to enable ease of integration with future solutions (Figure 14.3).

Digital Technology and Solutions Addressing Critical Pain Points in the OR

An intelligent hospital focuses on three major areas – operational efficiency, clinical excellence, and patient centricity, with technological advances being leveraged for these three areas to derive insights and improve performance. To support clinical excellence, solutions must be implemented for nurses and doctors to perform their tasks efficiently. Apart from using state-of-the-art medical equipment, departments of general surgery, emergency, radiology, intensive care, and outpatient departments also need to implement solutions to make processes faster and more streamlined, achieving the best possible patient outcomes.

Effective surgical setting solutions in hospitals must also provide patient-centric services that are not necessarily aligned to the clinical outcomes. For example, smart patient rooms, enable services that enhance patient well-being during their hospital stay, and provides aoverall hospital design that reduces the fear of being in a hospital which is crucial to have satisfied patients.

Labor makes up the largest portion of variable costs, hence methods of compensation and scheduling are critical for efficient utilization of labor. It is important to ensure that the number and distribution of staff match the workload. Hence when scheduling cases in an OR, it is critical to address the duration of the cases and to the number of working hours expected of the staff. Both under and over utilized time can have negative results.

Building Blocks Promoting Efficiency in the OR

When it comes to effective OR management, it's all about maximizing efficiency in order to realize cost savings, while ensuring maintenance and enhancement of quality of care. An OR with appropriate tracking and management systems will help in measuring and managing the underlying cost structure. Identifying the most efficient and optimal paths for the movement of resources is the ultimate aim of the operational team in a hospital. However, an OR is a difficult and complex environment, with inconsistent timing of input and output data, and a high number of constraints and variables:

- Variability in terms of the patient needs and the care to be provided.
- Variability of the resources to be used and time allocation
- Variations in the planning and the scheduling of care delivery

Critical Pain Points

The adoption of standard pathways and clinical processes is a necessary evolution. However, the vast spectrum of procedure types, the unpredictability of the duration of interventions, and taking into account both last-minute changes and the prioritization of patients on the waiting list, challenge most organizations in defining a more predictive practice model. The planning and scheduling of interventions within an OR area is a complex process that starts upstream with the management of a waiting list with patients queuing until a date, and a time is allocated in an OR theater. Another challenge lies in terms of the selection of patients on the waiting list that would need to be prioritized based on the surgery type and complexity so as to be served first.

OR scheduling is challenging, because of the complexities involved in the preoperative, intra-operative, and postoperative stages. In most hospitals, when OR blocks are assigned to surgery groups, there is no specific mechanism to ensure the availability of downstream resources. Hence patients cannot be sent to the next stage, but are held in the current stage, causing delays and backups. Blockages negatively impact OR management across the period process, such as increased waiting time and length of stay.

A growing number of hospitals today are turning to block scheduling to address improved OR management. Block scheduling is the process of allocating OR resources to a surgeon or group of surgeons for a specified day and time. It is a way to better manage OR time, which is an OR's most valuable resource.

Apart from being a more efficient use of time, the same surgeon or service line with subsequent cases results in fewer equipment and instrument changes and room positioning adjustments. The staff and surgeon establish a routine, which can be repeated through subsequent cases. Consistent access

Overview of Operating Room and Surgical Volumes

Integrated ORs attracts high-caliber medical practitioners while also meeting the needs of rising procedure volumes.

With 60 to 70% utilization of ORs permitting an average of 2 to 4 procedures per OR per day, ORs are positions to be the primary platforms for digitally enabled health care

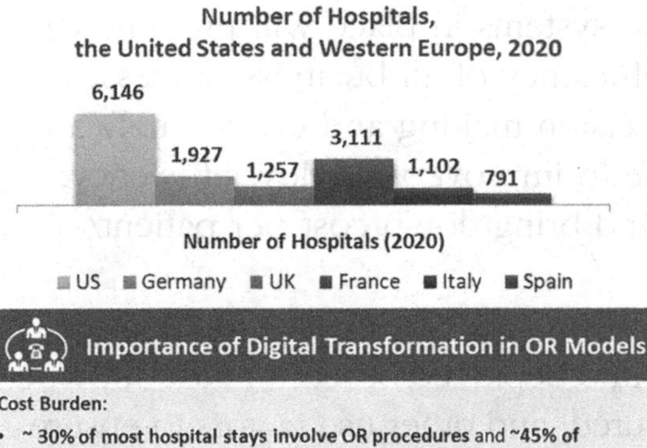

Digital Operating Room: Procedure Volumes, the United States and Western Europe, 2020

Country	Nos. of OR	Nos. of Operations/Year, 2020 (Estimated in Millions)
• United States	~40,664	~48.3
• Germany	~7,943	~10.7
• United Kingdom	~7,200	~10.3
• France	~4,142	~11.5
• Italy	~3,100	~5.3
• Spain	~2,723	~4.9

Importance of Digital Transformation in OR Models

Cost Burden:

- **~ 30% of most hospital stays involve OR procedures** and **~45% of the hospital costs** were attributed to stays that were a result of OR procedures.
- Market analyses and studies quote that a minute spent within an OR for a patient is a cost equivalent of **$150–250.**

Efficacy and Accuracy in Healthcare:

- **~ 3.5-4 billion patients** globally lack access to safe and affordable health care services and an estimated **~1 billion** patients lack access for surgical care.
- In Germany, France, and the United Kingdom, the most common specialization tends to be generalist medical practitioners.

Future for ORs with Digital Transformation

- **Augmented reality** to become mainstream in most hospital ORs, which will benefit manufacturers, providers, payers, and patients.
- **IT Integration** to enhance optimization and capabilities with hybrid OR models with AI, robotics, visualization, and data analytics. **Vendors that offers a full range of OR products, allows them to bundle their products.** This decreases the average sales price for their components and solidifies their market share.

Figure 14.4 Overview of operating room and surgical volumes.

Source: Frost & Sullivan

to the OR can be an incentive when recruiting new surgeons and a motivator for existing surgeons to maintain their case volumes in an effort to meet established block utilization targets.

The most significant benefit of block scheduling is increased utilization of OR time, which improves the operational efficiency and financial performance of an OR. The OR schedule becomes more predictable, which leads to less surgeon and patient waiting time and higher satisfaction. This standardization also helps hospitals to understand the basic volume guarantee and hence enables centers to plan equipment, instrumentation, and staffing based on this guaranteed business. The only real drawbacks to block scheduling occur when it is poorly managed, and blocks are underutilized (Figure 14.4).

Digital Technologies and Solutions Addressing Complexity and Process

Digital technologies are commonly focused on people and processes:

- OR Resources: The main aim being to bring about process adherence from all levels of staff aligned with organizational goals. Improved resource productivity can be achieved with the aid of personal incentives to help achieve target metrics.
- Process Management: Streamlined systems in place will provide real-time feedback and data on the efficiency of all business processes. Outputs can be used to guide decision making and enact timely adjustments. The main focus will be to improve workflow efficiency, reduce time and manual effort, and bring down cost per patient/procedure (Figure 14.5).

Technology will need to be evaluated and deployed based on cost of ownership, revenue generation, savings captured, and other use factors to ensure support to people and processes and to help identify new revenue streams.

Performing a successful surgical procedure is about art, skills, and precision, which can be supported and advanced by the latest technologies. Today, technology has evolved into user-friendly, powerful platforms that cover a number of aspects of surgery, including preoperative planning, component rotation, quantification of soft tissue balance, custom-made gigs, and robotic surgery for precise prosthetic fit.

Advanced computers and technology play an important role that can help increase accuracy and precision in the OR. With preoperative CT scans and

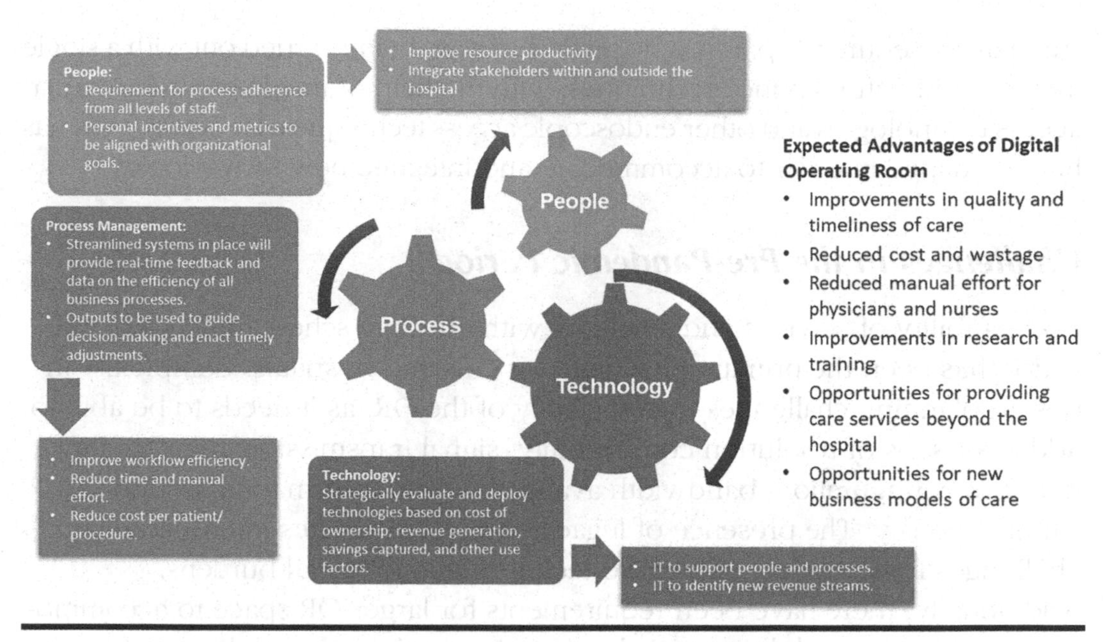

People:
Requirement for process adherence from all levels of staff.
Personal incentives and metrics to be aligned with organizational goals.
Improve resource productivity
Integrate stakeholders within and outside the hospital
Process Management:
Streamlined systems in place will provide real-time feedback and data on the efficiency of all business processes.
Outputs to be used to guide decision-making and enact timely adjustments.
People
Process
Technology
Improve workflow efficiency.
Reduce time and manual effort.
Reduce cost per patient/ procedure.
Technology:
Strategically evaluate and deploy technologies based on cost of ownership, revenue generation, savings captured, and other use factors.
IT to support people and processes.
IT to identify new revenue streams.
Expected Advantages of Digital Operating Room
Improvements in quality and timeliness of care
Reduced cost and wastage
Reduced manual effort for physicians and nurses
Improvements in research and training
Opportunities for providing care services beyond the hospital
Opportunities for new business models of care

Figure 14.5 Digital operating room – complexity and process.
Source: Frost & Sullivan

MRIs, planning can be done in three dimensions (3D). Under these circumstances, the surgeons take a deeper look into the bone shape, size, and quality. The images can be rotated and adjusted so as to identify and obtain the best possible fit for patients. One of the major advantages being that the OR team can plan surgeries days and sometimes weeks before arrival in the OR, allowing them to anticipate the surgery, having implants and special equipment ready to be used. This minimizes assumptions during surgery and makes outcomes more precise and predictable. New age and minimally invasive technologies strive to make surgeries more precise as well as aid to diminish other related problems of surgery, such as infection, bleeding, and post-operative pain.

Status of the OR

New technologies and devices in the OR are being introduced into a technologically complex environment, with the goal of improving health outcomes as well as OR efficiency, which in turn impacts patient safety and cost of care. The present surgical ecosystem is migrating from an invasive to less invasive and even noninvasive procedures, as minimally invasive surgery (MIS), image-guided procedures, robotic surgery, and tele-surgery continue to replace traditional surgical procedures. Laparoscopic procedures are being carried out with a single incision and natural orifice techniques, with the help of image-guided vascular access technologies and other endoscopic access techniques. Hence, the OR has had to change in order to accommodate and integrate new technologies.

Challenges in the Pre-Pandemic Period

Compatibility of services and solutions with effective scheduling and optimization has been the primary challenge for ORs and hospitals. Complexity in design planning challenges the efficiency of the OR, as it needs to be able to address issues of resolution compatibility, signal transmission compatibility, network configuration, bandwidth availability, and information security simultaneously. The presence of legacy equipment creates interoperability challenges for providers, adding to technical and financial burdens. Additionally, there have been requirements for larger OR space to accommodate additional capabilities or technology integration now and in the future.

The drive to enhance efficiency and increase surgical output per facility will drive the uptake of digital ORs. A digital OR aims to integrate images, documents, and workflow in the OR, inclusive of distributing video and images to the OR and displaying images from different sources to a single screen. Surgical procedures are also transitioning to an ambulatory setting, which will decrease the need for the conventional multidisciplinary OR and pave the way for specialized image-guided procedure suites and digital ORs (Figure 14.6).

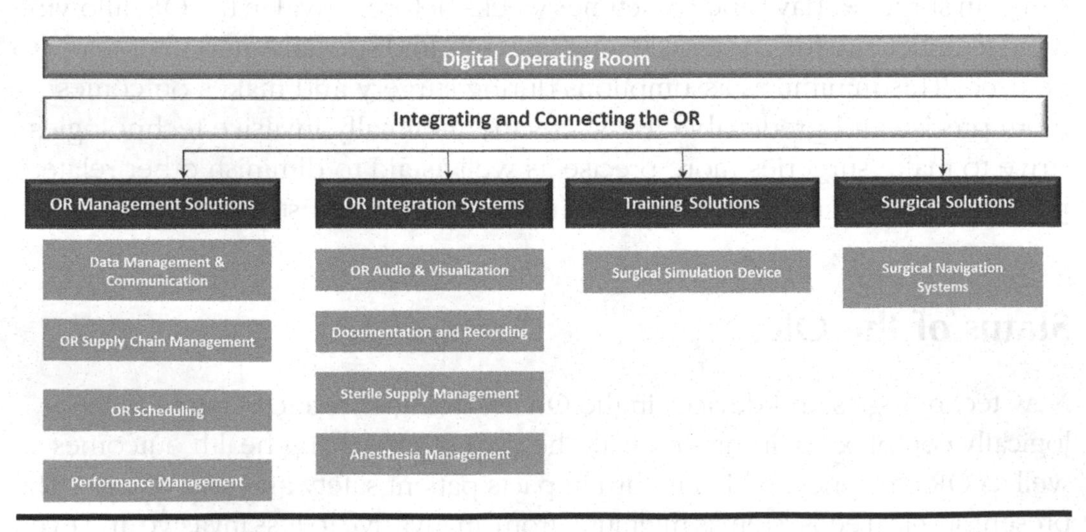

Figure 14.6 The digital operating room.

Source: Frost & Sullivan

Future Trends of the OR

The future of the OR will need to focus on flexibility as technologies will change and hospitals and surgical centers will need to embrace them. Modular accommodation with movable ceilings and walls. HVAC is primed for expansion and the provision of more than adequate power and air exchange rate in the room to make the most of technology.

ORs are beginning to prefer an open architecture in response to the growing complexity. Practices in healthcare are collaborating with architecture firms, research groups, etc., as seen in the Pebble Project and Realizing Improved Patient Care through Human Centered Design in the Operating Room (RIPCHD.OR) to improve safety and efficiency in the OR.

Data is becoming even more critical for hospitals, to measure success, provide insights to reduce risk, and support efforts to improve efficiency. Increasing complexity requires OR managers to ensure workflow balance through the course of the day with ease of control and flexibility. Simplified OR processes yield increased productivity and optimization. As payers push for standardization and value in the healthcare system, the benefits of smart and digital ORs will be realized.

As hospitals aim to rationalize their costs, updated systems enable clearer communication exchanges, effective responsiveness through improved staff monitoring, and standardization with inventory management. Uncertainty is a major problem associated with OR scheduling and surgical planning. Digital solutions offer opportunities for automation and solutions that help with planning and implementing the cost-effective management of resources for an OR.

Near-Term Trends in the OR

Navigation and positioning with minimal use of radiation and a high degree of minimal invasiveness would ensure a reduction in costs and shorter hospital stays and quicker recovery for patients.

The future would be through the use of sensors wherein an IP interface would enable connection to one IP hub. A connected ecosystem with sensors and medical devices will help serve the function of capturing and measuring data, identifying and stratifying possible risks. This in turn will enable the healthcare team to make informed decisions to enable them to take planned steps proactively.

Devices will work to form a large network ecosystem wherein data is exchanged to improve collaborative work. Open, standardized communication will help hospitals and healthcare centers maintain efficient data exchanges to and from the ORs across departments, hospitals, service providers, regions, and even countries.

A deep IT integration will be a required setup to reduce complexities and simplify workflows. The main aim will be to target efficiency by eliminating cumbersome processes and enabling gesture control and speech control for effective handling in the sterile field. Comparison of various documented surgical videos and results can be used to determine the most effective measures to ensure optimal results with minimal complications and lower percentages of post-surgical complications.

Robots and assist devices will help combat the challenge of physician fatigue and enable precision movements. The urgent need is to ensure that the surgeon has better control of the OR space without the need to deviate from the patient field. A sterile OR becomes even more critical to ensure patients are safe from life-threatening pathogens. OR hands-free technologies such as voice and gesture will enable surgeons to ensure better outcomes and patient safety.

Augmented reality (AR), along with 3D virtual reality (VR), surgical guidance, and navigation in major OR procedures would help ensure precision during surgery with pre-surgical hologram visualization of anatomical structures. This would identify possible hindrances and challenges that can be anticipated in order to preplan and alert the team and help them achieve optimal results.

Surgical simulation will provide safe and standardized training modules to permit learning from previous errors and help build confidence in surgeons. Surgeons will be able to practice the surgery process on three-dimensional models to ensure familiarity with the process to be executed on a specific patient in the OR. VR-based simulation enables experienced and trainee surgeons to build skill sets through gamification and life-like scenarios.

However, learning to conduct a specialized procedure takes approximately 6 months to a year, making this very expensive. Software algorithms are being used as they are capable of calculating an instrument's proximity to body tissues on a virtual platform. This model helps surgeons understand virtually how various tissues react and behave when cut or punctured during interventional procedures.

New-age surgeons will prefer to rely on haptic technology for a range of things from routine diagnosis to complex surgical procedures, hence making simulators capable of creating close to real-life scenarios for more effective training. Alerts, in the form of haptic responses, will further help surgeons recognize if the pressure threshold is reached, which, in turn, helps them prevent damage to anatomical structures.

The healthcare industry will become one of the first to acknowledge the impact of augmented and virtual reality technologies on human behavior, patient experience, and ultimately, in saving lives. It is estimated that by 2025, patients and consumers will always have a snapshot of how their bodies will be impacted by time and unhealthy lifestyles. Health conditions related to vision, cardiovascular and lung disorders, mental health, etc. will all be thoroughly researched and analyzed in real-time through AR-based applications, thus helping and assisting the patient/consumer in his/her overall well-being (Figure 14.7).

Topic	Present	Future Outlook
Data Collection & Storage	• Data transfer via the Internet forms the basis for all smart developments. • IP is being widely used for voice and video over IP and for information exchange between devices through IoT.	• The future of all devices would be through the use of sensors wherein an IP interface would enable connection to one IP hub. • Devices will work to form a large network ecosystem wherein data is exchanged to improve collaborative work. • Open, standardized communication will help hospitals and healthcare centers maintain efficient data exchanges to and from the ORs across departments, hospitals, service providers, regions, and even countries.
Productivity & ROI	• Catering to a larger number of patients is presently the mode of increasing productivity and ROI in all service centers. • Productivity is influenced and driven by competent workflow solutions and roistering software for efficient and timely task allocation.	• A deep IT integration will be a required setup to reduce complexities and simplify workflows. • The main aim will be to target efficiency by eliminating cumbersome processes and enabling gesture control and speech control for effective handling in the sterile field. • Comparison of various documented surgical videos and results can be used to determine the most effective measures to ensure optimal results with minimal complications and lower percentages of post-surgical recalls.
Clinical Surgery	• **Robot-assisted minimal invasive surgeries offering greater precision with better imaging. The main domains include cardiac surgery, cardiology, neurosurgery, gynecological oncology, and transplantation.**	• **AR, along with 3D virtual reality, surgical guidance, and navigation to ensure precision during surgery with pre-surgical hologram visualization of anatomical structures.** • **This would identify possible hindrances and challenges that can be anticipated in order to preplan and alert the team and help them achieve optimal results**
Workflow Solutions	• The OR environment lays great emphasis on safety and efficiency. Technology is adopted to reduce the OR team size and resource turn-around time and expedite the OR setup.	• Increasing demand for identifying population indices and routing patient traffic to enable specific services for primary, secondary, and tertiary care centers. • Enabling geographic mapping of disease population and identification of disease-free targets using patient tracking systems will help ensure demand-supply ratio. This also enables the hospital environment to attain a higher level of automation.
Precision Control Dependence on Surgeon	• Robots and assist devices can help combat the challenge of physician fatigue and enable precision movements. The urgent need now is to ensure that the surgeon has better control of the OR space without the need to deviate from the patient field.	• A sterile OR becomes even more critical to ensure patients are safe from life-threatening pathogens. OR hands-free technologies such as voice and gesture will enable surgeons to ensure better outcomes and patient safety.

Figure 14.7 Trends for future OR.

Source: Frost & Sullivan

What Does the Ideal OR Look Like in the Future?

The long-term future will hopefully be surgical procedures with minimal staff or even without a doctor or nurse. With surgical robots taking on most of the major procedures, connectivity would be the most sought after technology to ensure control of assisted devices from remote settings. Services and solutions would be focused on ensuring each OR is future-proof in terms of design and future requirements (Figure 14.8).

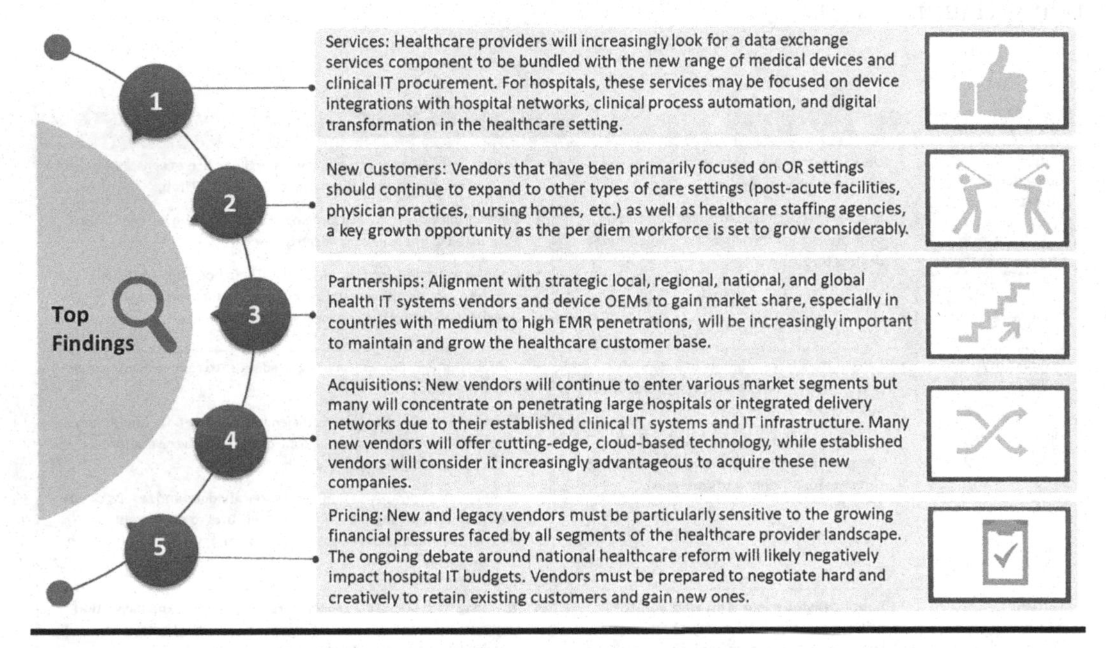

Figure 14.8 Strategic imperatives for growth.

Source: Frost & Sullivan

Chapter 15

Techquity

Chris Landon

In 1895, poet Joseph Malins (1) recognized "To rescue the fallen is good, but t'is best: To prevent other people from falling; Better to put a strong fence'round the top of the cliff: then an ambulance down in the Valley."

In my nearly five decades of caring for children with cystic fibrosis, the science supported by the Cystic Fibrosis Foundation resulted in one of the first possibilities of personalized genomic medicine. Over a decade ago, Joanna Fanos (2,3) outlined the difficulties for these "first families," children with diseases diagnosed at birth with new genetic testing, with chronic and grim prognoses, and cures yet undiscovered. She had watched her own sibling with cystic fibrosis, the relentless progression of its pulmonary disease, and its effect on her family. Parents struggled to get to center care, follow complex regimens, and "fighting for a cure." For more than 90% of the greater than 30,000 patients with cystic fibrosis in this country, there is a "cure."

What we now face, however, are the "last families," for whom there is no magic pill, who will continue to struggle. These "last families," with newborn genetic screening identifying them from birth as "non-responders" to highly effective modulator therapy, are predominantly Hispanic and Black leading to persistent inequities and health disparity among this group of people with CF as well.

The Cystic Fibrosis Foundation Registry now has to describe a different course for responders, non-responders, and lung transplants. The need to "fight for a cure" has been diminished for the majority of the patients and their families.

In a previous study of LatinX and BlackX in New York City, there was no difference in number of visits to the CF clinic, yet pulmonary function and

DOI: 10.4324/9781032690315-15

disease course differed in the rate of decline (4,5). There may be contributing individual factors such as adherence to treatment regimens, self-management culture, health literacy, and English proficiency that affect outcomes. The importance of designing culturally appropriate preventative and management strategies to better understand how to direct interventions to this vulnerable CF population is paramount.

The development of Cystic Fibrosis Foundation Guidelines (6,7) devoted to pulmonary and nutritional care, expanding to diabetes care and bowel cancers as the life span of the CF patient was extended, and referrals for lung transplant are in the process of change. For the 90% of highly effective modulator therapies, the need to consume their family's day with early morning awakenings before school for inhaled Pulmozyme, hypertonic saline, high-frequency chest wall oscillation, and missed school due to hospitalization has been radically altered. The need for the quarterly visits to the CF Center teams of nurses, dietitians, physical therapists, social workers, nurses, and specialty physicians will be altered as well, except for the need to renew the highly effective modulator therapy. Geriatric CF Clinics will begin to emerge. These treatments and visits have not changed for the predominantly Hispanic and Black "non-responders." In our previous work looking at studies of pediatric intensive care, studies were often anecdotal and included small sample sizes. Methodologic limitations were numerous and varied and seriously narrowed the significance of the studies we reviewed. The reports that we evaluated were largely limited to those of English-speaking families, white people, and married mothers.

Southern California Cystic Fibrosis Collaborative (SCFCC) was recently formed to facilitate communication and referral of study patients between CF Centers. Three additional subcommittees were formed for research, lung transplant, and health equity. As an initial step, SCCFC members underwent a survey to assess basic knowledge and practices in diversity, equity, and inclusion (DEI). Survey was sent to email addresses of members (91 in total) listed from seven centers. Questions: did they have a Health Equity Program within their institutions; what internal/external DEI resources were available; and if members of the institution would like to set up a time to discuss health equities within their institutions (Figure 15.1).

A total of 14 of the 91 surveys were returned. Of the 14 submitted surveys, 5 of the surveys indicated that members of their respective institutions stated they had a Health Equities Program. In total, 3 institutions use training modules, 2 use consultations, 1 uses data tracking and analysis for quality/equity improvement, 1 uses advisory councils/reports, and 3 use assistance

Cystic Fibrosis Conference: Health Equity Survey

Q3 If yes, what internal resources (for providers and staff) are available to help minority populations receive equitable treatment and care for Cystic Fibrosis? Please check all that apply.

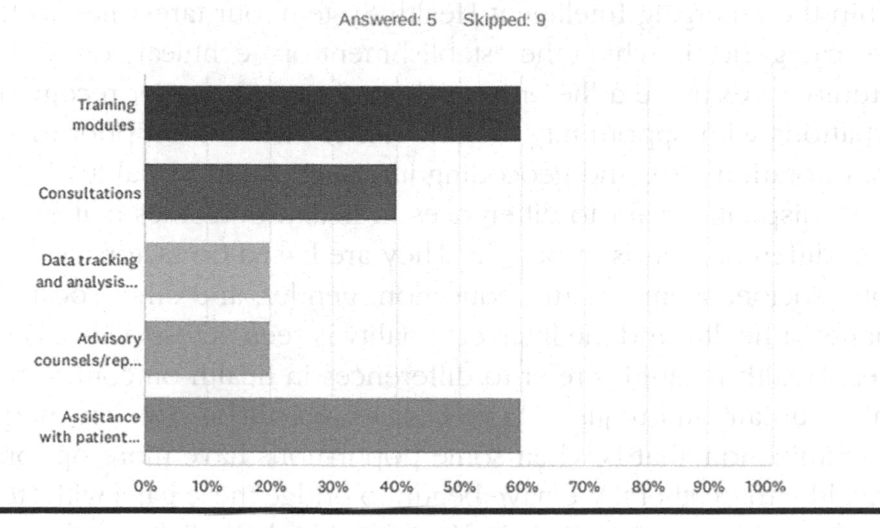

Figure 15.1 Internal resources provided to assist in health equities according to survey results.

with patient outreach programming. In terms of external resources, 12 of the 14 individuals responded. In total, 4 institutions have social assistance, 7 have financial, 6 have transportation, 4 have legal, 11 have language, 5 allow for opportunities to serve on advisory councils, and 3 have other assistance programs. When asked whether members would like to set up a meeting to discuss health equities, 13 of the 14 members agreed and 1 declined. Of those who wanted to set up a meeting, 9 of the 13 responded with dates and times. One comment that stood out goes as follows, "We felt that we were on task with our minority patients, even those who do not have English as their primary language. Then, our SW (who will be answering also) stated that even though we give our non-English speaking families all the assistance they need to obtain insurance, have outside resources available to them, obtain medications we actually do not 'socialize' with them like we do our English-speaking families and I realized that is so true. Non-English speaking families' children may become more comfortable with English and we may ask them simple questions, but we found that their parents are sometimes hesitant to express themselves for fear of embarrassment, being looked at as asking 'dumb' questions, and just not wanting to be a bother. Just an observation that we actually were discussing today."

LatinX and BlackX Cystic Fibrosis patients are at greater likelihood of being non-responsive to HEMT – the clinic population requiring frequent patient visits will be shifting significantly to patients with socioeconomic risk factors impacting their disease course.

Within the emerging Intelligent Health System, our target lies within the last two categories in which the establishment of "technical equity" is the opportunity to examine adherence to therapies (8,9), earlier recognition of exacerbations with opportunity to seek earlier intervention prior to perceivable deterioration (10), and geocoding intersectionality variables (11).

Health disparities refer to differences in health outcomes that exist between different groups of people. They are based on factors such as race, ethnicity, socioeconomic status, education, gender, and other social determinants of health, and the intersectionality is seen across heart disease and diabetes. Health inequities refer to differences in health outcomes that are avoidable, unfair, and unjust. They occur as a result of systemic inequalities and discrimination, that is when some populations have more opportunities to be healthy than others. We have begun to bridge these gaps with "technical equity" through the use of COVID-19-enforced telemedicine and remote patient monitoring (12). A potential solution could be the use of a patient-personalized model system in these LatinX and BlackX people with CF to transform data from bench to bedside to identify patient-specific therapies and responsiveness to HEMT that we have reported previously (13). Application of the evolution of these products through machine learning, skin color appropriate pulse oximetry, home sleep testing in the chaotic urban environment, geocoding, and connecting these dots through healthcare application of distributed ledger technology may help us to build a better fence at the top of the cliff.

References

[1] Malins, J The Ambulance Down in the Valley. 1895

[2] Fanos, J New "first families": the psychosocial impact of new genetic technologies. *Genet Med.* 2012;14:189–190. 10.1038/gim.2011.17

[3] Oates GR, Schechter MS Social inequities and cystic fibrosis outcomes: We can do better. *Ann Am Thorac Soc.* 2021 Feb;18(2):215–217.

[4] DiMango, E, Simpson, K, Menten, E *et al.* Health Disparities among adults cared for at an urban cystic fibrosis program. *Orphanet J Rare Dis.* 2021;16:332.

[5] Buu MC, Sanders LM, Mayo JA, Milla CE, Wise PH Assessing differences in mortality rates and risk factors between hispanic and non-hispanic patients with cystic fibrosis in California. *Chest*. 2016;149(2):380–389.

[6] Flume PA, Mogayzel PJ Jr, Robinson KA, Goss CH, Rosenblatt RL, Kuhn RJ, Marshall BC Clinical Practice Guidelines for Pulmonary Therapies Committee. Cystic fibrosis pulmonary guidelines: Treatment of pulmonary exacerbations. *Am J Respir Crit Care Med*. 2009 Nov 1;180(9):802–808. doi: 10.1164/rccm.200812-1845PP. Epub 2009 Sep 3

[7] Rho J, Ahn C, Gao A, Sawicki GS, Keller A, Jain R Disparities in mortality of hispanic patients with cystic fibrosis in the United States. A National and Regional Cohort Study. *Am J Respir Crit Care Med*. 2018 Oct 15;198(8):1055–1063. doi: 10.1164/rccm.201711-2357OC. PMID: 29742360; PMCID: PMC6221571.

[8] Landon C, Papador J, Turner R Fostering adherence to help exacerbation reduction – tool for health coaching in cystic fibrosis. *Am J Respir Crit Care Med*. 2017;195:A5338

[9] Shudy M, de Almeida ML, Ly S, Landon C, Groft S, Jenkins TL, Nicholson CE Impact of pediatric critical illness and injury on families: a systematic literature review. *Pediatrics*. 2006 Dec;118 (Suppl 3):S203–S218. doi: 10.1542/peds.2006-0951B. PMID: 17142557.

[10] Swaminathan S, Toros B, Wysham N, Mark N, Ramanathan S, Morill J, Ka=onda V, Iyas S, Landon C Vironix: remote screening, detection, and triage of viral respiratory illness via cloud-enabled, machine-learned APIs. *Eur Respir J*. 2021:58. DOI:10.1183/13993003

[11] Vo J, Carroll N, Landon C Wildfire smoke in strawberry fields: evaluating fire-associated PM 2.5 exposure effects on Ventura County farmworkers. *Chest*. October 2021;160(4), Supplement, A1928.

[12] Landon C COVID-19 Wellness Monitoring Turns To Health Monitoring. *Closing The Care Gap With Wearable Devices*. 2023:143–152.

[13] Arora K, Yang F, Brewington J, McPhail G, Cortez AR, Sundaram N, Ramananda Y, Ogden H, Helmrath M, Clancy JP, Naren AP Patient personalized translational tools in cystic fibrosis to transform data from bench to bed-side and back. *Am J Physiol Gastrointest Liver Physiol*. 2021 Jun 1;320(6):G1123–G1130. doi: 10.1152/ajpgi.00095.2021. Epub 2021 May 5.

Appendix I: Glossary of Terms

3D Printing: 3D printing, or additive manufacturing, is a way of making objects by depositing layers of material based on a digital model.

802.11 Standard: A wireless local area network (WLAN) standard in the 2.4, 3.6, and 5 GHz frequency bands. It is maintained by the IEEE LAN/MAN Standards Committee (IEEE 802). 802.11 Standard is the basis for Wi-Fi specification. This technology is used by some RTLS solutions to calculate location.

802.15.4 Standard: A standard which specifies the physical layer and media access control for low-rate wireless personal area networks (LR-WPANs). It is maintained by the IEEE 802.15 working group. 802.15.4 Standard is the basis for the ZigBee specification. This technology is also used by some RTLS solutions to calculate location.

Accountable Care: A healthcare delivery and payment model that ties provider reimbursement to quality metrics and reductions in total cost of care. The Accountable Care Organization (ACO) is a group of coordinated healthcare providers that provides care to a given population.

Active RFID Systems: In active RFID systems, tags have their own transmitter and power source, usually a battery. Active tags broadcast their own signal to transmit the information stored on their microchips. Active RFID systems typically operate in the UHF band and offer a range of up to 100 m. In general, active tags are used on large objects, such as rail cars, big reusable containers, and other assets that need to be tracked over long distances. They are more expensive than passive tags.

Additive Manufacturing: Additive manufacturing, or 3D Printing is a way of making objects by depositing layers of material based on a digital model.

Artificial Intelligence: The ability of a digital computer or computer-controlled to perform tasks commonly associated with intelligent beings. The term is frequently applied to the project of developing systems endowed with the intellectual processes characteristic of humans, such as the ability to reason, discover meaning, generalize, or learn from past experience.

Application: Software that utilizes data coming from middleware and that directly interacts with the end user. Examples of applications include asset tracking and management, patient flow, temperature monitoring, infection control and hand hygiene, staff duress, inventory tracking and management, positive patient identification, business intelligence and reporting, and wireless nurse call.

Air Interface Protocol: A radio-based communication link protocol that governs how tags and readers communicate.

Augmented Reality: Augmented reality (AR) is an interactive experience that combines the real world and computer-generated content. The content can span multiple sensory modalities, including visual, auditory, haptic, somatosensory, and olfactory. AR can be defined as a system that incorporates three basic features: a combination of real and virtual worlds, real-time interaction, and accurate 3D registration of virtual and real objects.

Backscatter: A reflection of waves back in the direction from which they came. RFID tags use backscatter technology to reflect radio waves back to the reader, usually at the same carrier frequency. The reflected signal is modulated to transmit data.

Bandwidth: The range or band of frequencies, defined within the electromagnetic spectrum, that a system is capable of receiving or delivering.

Barcodes: Barcodes consist of small images of lines (bars) and spaces affixed to retail store items, ID cards, and postal mail to identify a particular product number, person, or location. A barcode reader uses a laser beam that is sensitive to the reflections from line and space thickness and variation. The reader translates information from the image to digital data and sends it to a computer for storage or for another process. 2D barcodes store information not only horizontally, as one-dimensional barcodes do, but vertically as well. That construction enables 2D codes to store up to 7,089 characters. The traditional, unidimensional barcode has only a 20-character capacity.

Battery-Assisted Passive Tag (BAP): RFID tags, with batteries, that communicate using the same backscatter technique used by passive tags (tags with no batteries). They use the battery to run the circuitry on the microchip and sometimes an onboard sensor. They have a longer read range than a regular passive tag because all of the energy gathered from the reader can be reflected back to it. They are sometimes called "semi-passive RFID tags."

Biometrics: Technology is used to identify individuals by comparing biological data, such as fingerprints, voice characteristics, and iris patterns, against stored data for that individual. Biometric systems consist of a reader or scanning device, software that converts the scanned biological data into a digital format and compares match points, and a database that stores the biometric data for comparison. Authentication by biometric verification is becoming increasingly common in corporate and public security systems, consumer electronics, and point of sale (POS) applications. Specific biometric AIDC (Automatic Identification and Data Capture) technologies include finger-scanning, electro-optical fingerprint recognition, finger vein ID, and voice recognition.

Bio Printing: Bioprinting (also known as 3D bioprinting) is the combination of 3D printing with biomaterials to replicate parts that imitate natural tissues, bones, and blood vessels in the body.

Biotechnology: Biotechnology is the use of Biology to solve problems and make useful products. The most prominent area of biotechnology is the production of therapeutic proteins and other drugs through genetic Engineering.

Bluetooth Low Energy (BLE): A wireless network protocol, especially suited for sensors and other small devices, requires extremely low power consumption. The protocol enables devices such as an iPhone to transmit a Bluetooth signal to beacons, with a read range up to 50 meters.

Connectivity: A generic term for connecting devices to each other in order to transfer data back and forth. It often refers to network connections, which embrace bridges, routers, switches, and gateways as well as backbone networks.

Coronavirus (COVID-19): Coronavirus disease 2019 is a contagious disease caused by severe acute respiratory syndrome coronavirus 2. The first case was identified in Wuhan, China, in December 2019. Symptoms of COVID-19 are variable but often include fever, cough, fatigue, breathing difficulties, and loss of smell and taste.

Cybersecurity: Cybersecurity is the practice of protecting critical systems and sensitive information from digital attacks. Also known as information technology (IT) security, cybersecurity measures are designed to combat threats against networked systems and applications, whether those threats originate from inside or outside of an organization.

Data Retention: The ability of a microchip to maintain the information stored in EEPROM (Electrically Erasable Programmable Read-Only Memory). RFID tags and other microchips can typically retain data for 10 years or more, but data retention depends on temperature, humidity, and other factors.

Environment of Care: The Environment of Care defines the overall patient care enviroment and can include environements for diagonostics, treatment and patient management surgical, radiation, critical care, imaging, laboratory, etc.). Each enveronment has specific building or space requirements to support optimized patient care, safety and satisifaction.

E-Pedigree or Electronic Pedigree: An electronic document provides data related to the history of a batch of a drug. The e-pedigree is used to identify a drug prior to sale, purchase, or trade of that product. The states of California and Florida have set deadlines for mandatory compliance to an e-pedigree system for tracking of medications through the supply chain. The FDA has also established electronic pedigree regulations to reduce the risk of counterfeit products.

Electromagnetic Interference (EMI): A phenomenon in which the electromagnetic field of one device disrupts, impedes, or degrades the electromagnetic field of another device, potentially blocking transmission.

Electromagnetic Spectrum: The range or continuum of electromagnetic radiation is characterized in terms of frequency or wavelenght.

Electronic Product Code (EPC): A unique identifier for every physical object anywhere in the world, for all time. Its structure is defined in the GS1 EPCglobal Tag Data Standard; an open standard available for download from the GS1 EPCglobal website. The EPC is designed as a flexible framework that can support a variety of existing coding schemes, including many coding schemes currently in use with barcode technology. EPC identifiers currently support seven identification keys from the GS1 system of identifiers. An EPC takes the form of a 96-bit string of data stored in a RAIN RFID tag chip that can act as a key to a database. The EPC is a Uniform Resource Number (URN) namespace registered to EPCglobal by the Internet Assigned Numbers Authority (IANA).

Encryption: Translation of data into a code for the purpose of keeping information secure from all but the intended recipient.

EPCglobal: A joint venture between GS1 (formerly known as EAN International) and GS1 US. It is an organization set up to achieve worldwide adoption and standardization of Electronic Product Code (EPC) technology.

Excite: The transmission of radio frequency energy from the RFID tag reader stimulates a passive RFID tag to provide power to transmit its data back.

Frequency: The number of cycles a periodic signal executes in unit time. It is usually expressed in Hertz (cycles per second) or appropriate weighted units such as kilohertz (kHz), Megahertz (MHz), and Gigahertz (GHz).

GS1: An international organization to develop and maintain standards for supply and demand chains. The four key standards of focus are Barcodes (used to automatically identify things), eCom (electronic business messaging standards allowing automatic electronic transmission of data), GDSN (Global Data Synchronization standards that allow business partners to have consistent item data in their systems at the same time), and EPCglobal (which uses RFID technology to immediately track an item).

Health Insurance Portability and Accountability Act (HIPAA): The US Congress enacted HIPAA in 1996 to regulate the interchange of private patient data to help prevent unlawful disclosure or release of medical information.

Healthcare Information and Management Systems Society (HIMSS): A not-for-profit association of 50,000 individual members and 570 corporate members. This organization is dedicated to improving healthcare quality, safety, access, and cost-effectiveness through the use of IT and management systems.

Healthcare Information Exchange (HIE): The process by which health information is mobilized electronically between or across organizations within a region, community, or hospital system. Federal and state regulations regarding HIEs are still being defined. In the meantime, multiple state and healthcare provider exchanges have been developed in the United States to manage movement of electronic records.

High Frequency (HF): The frequency bandwidth is from 3 MHz to 30 MHz. HF RFID tags typically operate at 13.56 MHz, can normally be read at short range (three feet or less), and transmit data faster than low-frequency tags, although they consume more power than low-frequency tags.

Infrared (IR): A technology that uses electromagnetic radiation with a wavelength that is longer than that of visible light, but shorter than that of microwaves and terahertz radiations. The IR signal does not penetrate walls, ceilings, floors, or large objects inside a room, but it does bounce off any object in its path. This technology is used to enable RTLS systems and is often used in conjunction with RFID.

Injectable Tech: Technology that is placed inside the body using an instrument to pierce the skin or mucous membranes.

Insertable Tech: Technology that is inserted in a natural orifice of the body (ear canal, tear duct, oral cavity, etc.).

Intelligent Hospital: The intelligent hospital is one that works better and smarter. It's better because it's resourceful, creative, and perceptive about what patients and doctors need, and it's smarter because it's astute and inventive when it comes to weaving together diverse technologies to enhance patient care.

Intensive Care Unit (ICU): A department within a hospital or healthcare facility dedicated to intensive care of patients with severe or life-threatening injuries or illnesses that require constant, close monitoring and support.

Interactive Voice Response (IVR): A telephony technology in which an individual uses a touch-tone phone to interact with a database to acquire information from, or enter data into, the database.

International Organization for Standardization (ISO): A non-governmental organization consisting of the national standards institutes of 205 countries. Each member country has one representative and the organization maintains a Central Secretariat in Geneva, Switzerland, that coordinates the system. Most RFID-related ISO Standards are: ISO 10536 – the international standard for proximity cards; ISO 11784 – the international standard defining frequencies, baud rate, bit coding, and data structures of the transponders used for animal identification; ISO 14443 – a set of international standards covering proximity smart cards; ISO 15693 – the international standard for vicinity smart cards; ISO 18000 – a series of international standards for the air interface protocol used in RFID systems for tagging goods within the supply chain; and ISO 7816 – a set of international standards covering the basic characteristics of smart cards, such as physical and electrical characteristics, communication protocols and others.

Interoperability: Interoperability is a characteristic of a product or system to work with other products or systems. While the term was initially defined for IT or systems engineering services to allow for information exchange.

Joint Commission on Accreditation of Healthcare Organization (JCAHO): A non-profit, US association that accredits healthcare organizations and programs as part of its mission to improve healthcare. A majority of state governments recognize the Joint Commission's accreditation as a condition of license and the receipt of Medicaid reimbursements.

LAN (Local Area Network): A relatively small network covering areas such as a room, a department, a building, a campus, etc.

Low Frequency (LF): The frequency bandwidth is from 30 kHz to 300 kHz. Low-frequency RFID tags typically operate at 125 kHz or 134 kHz. Low-frequency RFID tags must be read from within three feet, and their data transfer rate is slow, but they are less susceptible to interference than UHF tags.

Machine to Machinc (M2M): The term for technologies that enable devices (such as sensors or meters) to communicate with each other or another device (such as an appliance). An M2M network can consist of connection between two devices or multiple devices. M2M healthcare applications include patient monitoring solutions or drug dispensing tracking.

Memory: A means of storing data in electronic form. A variety of random access (RAM), read-only (ROM), Write Once Read Many (WORM), and read/write (RW) memory devices can be used. In RFID terms, memory is the amount of data that can be stored on the microchip in an RFID tag. It can range from 64 bits to 2 kilobytes or more on passive tags.

Mesh Network: A Mesh network consists of wireless nodes that relay data to other nodes as part of a network that ultimately forwards data to a server. An example of a mesh network is Zigbee digital radio technology that forwards data via "hops" from one node to another.

Micro-Electrical Mechanical Systems (MEMS): A term that refers to the combining of electrical and mechanical components on a chip to produce a very small system. MEMS medical devices are being developed to replace traditional devices because they are so small they can be used within a patient's body or a very small tool such as a scalpel, for example, in surgery. The MEMS devices promise to be more sensitive and robust than traditional technology. MEMS technology can be used within a host body or in biological samples to detect and diagnose health status.

Microwave Tags: A term that is sometimes used to refer to RFID tags that operate at 5.8 GHz. They have very high transfer rates and can be read from as far as 30 feet away, but they use a lot of power and are expensive. (Some people refer to any tag that operates above about 415 MHz as a microwave tag.)

Middleware: In the RFID world, this term refers to software that resides on a server between readers and enterprise applications. The middleware is used to filter data and pass on only useful information to enterprise applications. Some middleware can also be used to manage readers on a network.

Modulation: A term to denote the process of superimposing (modulating) channel encoded data or signals onto a radio frequency carrier to enable the data to be effectively coupled or propagated across an air interface. Modulation is also used as an associative term for methods used to modulate carrier waves. Methods generally rely on the variation of key parameter values of amplitude, frequency, or phase. Digital modulation methods principally feature amplitude shift keying (ASK), frequency shift keying (FSK), phase shift keying (PSK), or variants.

Near Field Communication (NFC): A set of standards for smartphones and tablets to establish radio communication with each other or tags by touching them or bringing them into close proximity, usually no more than a few centimeters from each other or a tag. Existing and anticipated applications include contactless transactions, data exchange, and simplified setup of more complex communications such as Wi-Fi.

Outcomes Measures: Used to assess the impact of health services in terms of improved quality and/or longevity of life and function.

Passive RFID Systems: In passive RFID systems, the reader sends a radio signal to the tag. The RFID tag then uses the transmitted signal to power on, and reflect energy back to the reader. Passive RFID systems can operate in the LF, HF, or UHF radio bands. Passive tags do not require a power source or transmitter and only require a tag chip and antenna. They are cheaper, smaller, and easier to manufacture than active tags and typically have a range of less than 15 m.

Patient Activation/Engagement: The extent to which patients have skills, knowledge, and motivation to participate as part of their care team.

Patient-Centric Identification and Association: Development of a standardized naming or identification methodology to enable the identification of a patient, staff, medical devices, supplies, and pharmaceuticals associated to a specific patient. This ensures specific processes commonly associated with the patient care and delivery are uniquely identified and associated for process management and workflows. As an example, alarm management and delivery (ensure delivery of alarms to the appropriate clinical staff), sample collection (associate blood samples, etc. to the correct patient), and medication administration (ensure delivery of the right drug to the right patient at the right time). This identification and association process minimizes errors and enhances patient safety.

Population Management: The health outcomes of a group of individuals, as well as the distribution of such outcomes within the group; an approach that aims to improve the health of an entire human population.

Protocol: A set of rules governing a particular function, such as the flow of data or information in a communication system.

Radio-Frequency Identification (RFID): Radio-frequency identification (RFID) is a form of wireless communication that uses radio waves to identify and track items. RFID can tell you what an object is, where it is, and even its condition, which is why it is integral to the development of the Internet of Things – a globally interconnected web of objects allowing the physical world itself to become an information system. RFID systems can be broken down by the frequency band within which they operate: low frequency (LF), high frequency (HF), and ultra-high frequency (UHF). There are also two broad categories of RFID systems – see Active RFID Systems and Passive RFID Systems.

RAIN RFID: RAIN RFID is a passive wireless technology system that connects billions of everyday items to the Internet, enabling businesses and consumers to identify, locate, authenticate, and engage each item. RAIN RFID is used in a wide variety of applications, including inventory management, patient safety, asset tracking, and item authentication. RAIN is the fastest-growing segment of the RFID market and uses a single, global standard: UHF Gen 2 (ISO/IEC 18000-63). The RAIN RFID industry is represented by a global alliance called the RAIN Alliance.

Read Only: The term applied to a transponder or tag in which data is stored in an unchangeable manner and can therefore only be read and not altered. Writing to a read-only tag is also impossible.

Read Range: The distance from which a reader can communicate with a tag. Active tags have a longer read range than passive tags because they use their own power source (usually a battery) to transmit signals to the reader. With passive tags, the read range is influenced by frequency, reader output power, antenna design, and method of powering up the tag. Low-frequency tags use inductive coupling, which requires the tag to be within a few feet of the reader.

Read Rate: The maximum rate at which data can be communicated between transponder and reader/interrogator, usually expressed in bits per second (bps or bits.s-1).

Read/Write: When applied to an RFID system, this term refers to the ability to both read data from a transponder and to change data (write data) using a suitable programming device. See RFID Reader.

Readmission: This term describes an admission to an acute care hospital within 30 days of discharge from an acute care hospital.

Remote Care or Virtual Care: Synonymous with Telemedicine or Telehealth, this term refers to the use of telecommunication and information technologies to provide clinical care at a distance.

Resolution: Determines how well a tagged person or item can be tracked to a specific location (e.g., resident room, ER bay).

RF-IR: A hybrid technology that uses both radio waves and infrared for identification and tracking purposes. This technology is used to enable some RTLS systems.

RFIDba: The International RFID Business Association (RFIDba) was founded in April 2004 as a not-for-profit, educational, technology, and frequency-agnostic, trade association dedicated to serving the business needs of the end-user community with vendor-neutral information on RFID and RTLS technologies along with information on other associated, complimentary technologies.

RFID in Healthcare Consortium: The RFID in Healthcare Consortium (RHCC) was founded on September 13th, 2008, initially to address EMI issues with RFID and RTLS technologies but has since transformed itself into a full-fledged trade organization. The Consortium is a primary source for vendor neutral, educational programs; industry information; and other value-added services pertinent to RFID and RTLS technologies. The RHCC serves the needs of end-user communities in the healthcare, assisted living, and nursing home industries.

Reader/Gateway (Connectivity): A RAIN RFID reader or gateway is a device that provides connectivity between tag data from an endpoint and the enterprise system software that needs the information. A reader connects to external antenna(s) while a gateway integrates the reader and antenna(s) into a single device. Readers and gateways communicate with tags (or endpoints) that are within their field of operation, performing any number of tasks including simple continuous inventorying, filtering (searching for tags that meet certain criteria), writing (or encoding) selected tags, etc.

RFID Receiver: A device that listens for RF (radio frequency) signals and converts them into data packets that are available for further processing. It is used with active RFID tags, which continuously emit pre-programmed messages.

RFID Transponder (Tag): A microchip attached to an antenna that can be applied to an object. The tag picks up signals from, and sends signals to, an RFID reader. The tag may contain a unique serial number and may have other information such as a customer account number. Tags come in many forms, such as smart labels that have a barcode printed on them, or the tag can simply be mounted inside a carton or embedded in plastic. RFID tags can be active, passive, or semi-passive.

Real-Time Locating System (RTLS): Refers to technology that is used to locate and track people and items (such as assets, equipment, and inventory) by associating a tag with each person or item. This term is commonly used in reference to "active," battery-powered locating technologies.

Robotic Surgery: Robotic surgery, also called robot-assisted surgery, allows doctors to perform many types of complex procedures with more precision, flexibility, and control than is possible with conventional techniques. Robotic surgery is usually associated with minimally invasive surgery—procedures performed through tiny incisions. It is also sometimes used in certain traditional open surgical procedures. The most widely used clinical robotic surgical system includes a camera arm and mechanical arms with surgical instruments attached to them. The surgeon controls the arms while seated at a computer console near the operating table. The console gives the surgeon a high-definition, magnified, 3D view of the surgical site. The surgeon leads other team members who assist during the operation.

Step-Down Unit: A hospital nursing area dedicated to the care of patients who have undergone surgery. As opposed to the ICU, the Step-down Unit serves patients who are stable but may require monitoring due to the recent surgery.

Swallowed Tech: Technology that is employed after swallowing.

Tag (Endpoint): A RAIN RFID tag is comprised of an integrated circuit (called an IC or chip) attached to an antenna that can be built into apparel hang tags, labels, wristbands, security tags, industrial asset tags, etc. The tag contains a unique identifier or EPC and when placed on an item creates an endpoint that communicates with readers to deliver the identity, location, and authenticity of items.

Telehealth: Sometimes called telemedicine. The use of telecommunications and IT to deliver health services and transmit health information to another location.

Telemetry: The automatic transmission and measurement of data from remote sources by wire, radio, or other means.

Transponder: An electronic transmitter/responder is commonly referred to as a tag.

Ubiquitous Sensor Network (USN): A network of sensors provides coverage of every single part of an area such as a healthcare facility, however remote. Such networks are typically managed from a control center.

Ultra-High-Frequency (UHF): The frequency bandwidth is from 300 MHz to 3 GHz. Typically, RFID tags that operate between 866 MHz and 960 MHz send information faster and farther than high- and low-frequency (HF and LF) tags. However, UHF signals cannot pass through items with high water content. UHF tags generally consume more power than low-frequency tags.

Ultrasound: A technology that uses a cyclic sound pressure with a frequency greater than the upper limit of human hearing. The production of ultrasound is used in Real-Time Location Systems (RTLS), as it can provide room-level location accuracy because the sound does not penetrate walls.

Ultra-Wideband (UWB): Any radio technology used at a very low energy level for short-range, short-duration, high bandwidth communications. UWB technology pulses within a bandwidth exceeding the lesser of 500 MHz or 20% of the center frequency. The UWB pulses are received by sensors which determine a tag's location based on Time-Difference-of-Arival (TDoA) and Angle of Arrival (AoA). This technology is used to enable accurate indoor positioning for RTLS systems.

Unique Device Identification (UDI): A system used to mark and identify medical devices within the healthcare supply chain. The FDA has released a rule that a unique number should be assigned by the device manufacturer to each version or model of a medical device, in both human-readable format and AutoID format (such as barcode or RFID).

Unique Patient Identifier (UPI): An identification code is used in the management of healthcare information and record-keeping. The UPI is used to identify and access patient care information as well as for medical record chart analysis, billing, and reimbursement. The Joint Commission on Accreditation of Healthcare Organization (JCAHO)'s Information Management Standards mandate that the unique patient identifier be part of a patient's medical records.

Video Conferencing: Also known as video-teleconferencing, this term refers to the use of a set of telecommunication technologies that allow two or more locations to communicate by simultaneous two-way video and audio transmissions.

Virtual Care: See Remote Care or Telehealth.

Wide Area Network (WAN): As its name suggests, this is a computer network that covers a far wider area than a LAN, such as cities, countries, continents, and the whole world. A WAN is formed by linking LANs together.

Wearable Tech: Technology that is worn on the body in the form of clothing, straps, hanging around the neck, socks, underwear, etc.

Wi-Fi: Refers to any system that uses the 802.11 Standard, which was developed by the Institute of Electrical and Electronics Engineers (IEEE). Wi-Fi networks operate in the 2.4 and 5 GHz radio bands, with some products that contain both bands (dual band). Wi-Fi is a very common wireless technology that is used to connect machines in a LAN. This technology is used by some RTLS systems for locating purposes.

Wireless Medical Telemetry Services (WMTS): The remote monitoring of a patient's physiological parameters including pulse and respiration rates using RFID or other medical telemetry devices.

Wireless Sensor Network (WSN): A network of spatially distributed autonomous sensors to cooperatively monitor physical or environmental conditions including motion, temperature, pressure, sound, or vibration. A sensor network is typically equipped with a radio transceiver or other wireless communications device, a small micro-controller, and an energy source – usually a battery.

Write: The RFID process of transferring data to a transponder (tag) from a reader, as well as storing the data on the transponder, may also encompass the reading of data to verify the data content.

Write Once Read Many (WORM): A label distinguishing a transponder that can be partially or totally programmed once by the user, and thereafter only read.

Write Rate: The rate at which data is transferred to a transponder and stored within the memory of the device and verified. The rate is usually expressed as the average number of bits or bytes per second over which the complete transfer is performed.

ZigBee: See 802.15.4 Standard.

Index

Note: *Italicized* and **bold** page numbers refer to figures and tables.

Abbott Diabetes, 73, 82
ACA. *See* Affordable Care Act
Accountable Care Organizations (ACOs), 173, 182
ACGIH. *See* American Conference of Governmental Industrial Hygienists
ACOs. *See* Accountable Care Organizations
Active RFID readers, 115–116, *116*
Active RFID tags, 123–125
Acute Hospital Care At Home program, 216, 221
Adhesives, for attachment, 85
Adibot, 100
Advanced communications, 23–24
Advanced medical systems, 17–19
Aethon TUG robot, 100
Affordable Care Act (ACA), 173, 220
AI. *See* Artificial intelligence
Airbnb, 6
Airway Trainer, 103
Alder Hey Children's Hospital, 101
Amazon
  Alexa, 2, 176
  Amazon Care, 6
  Amazon Go, 7
Ambulatory care
  background of, 162
  challenges to, 162
  contract tracing, 167–168, *168*
  impact on patient volumes and revenues, 163
  impact on workflow and operations, 163–164
  impact on workforce, 163
  patient throughput, improving, 166–167
  physical waiting room, elimination of, 165
  real-time location systems, role of, 162–169
  real-time tools, 164–165
  solution, 164
  staff workflow, improving, 166–167
  staying connected to families and visitors, 168–169
  streamlined door-to-exam patient flow, 165, *166*
  virtual waiting rooms, creation of, 165
Amedisys, 222
American Conference of Governmental Industrial Hygienists (ACGIH), 192
American Society of Heating, Refrigerating, and Air-Conditioning Engineers (ASHRAE), 198
Anatomical replicas, 103–104
Apple Watch, 74–75, 83
AR. *See* Augmented reality
Arthrobot, 87, 90
Artificial intelligence (AI), 1–2, 7, 75, 174–177, 179, 181, 182
  conversational, 10
  guided assistance of robotics, 95
  IoHT and, 48–49
  use cases, 48–49
ASHRAE. *See* American Society of Heating, Refrigerating, and Air-Conditioning Engineers

Asset
  definition of, 24
  inventory, 55–59
  management, 55–59, 132, 136
  tagging, 136
  tracing, 133–134
  tracking, 133–134, 137
  visualization, 27, *28*
Atrium Health, 225
Atrius Health, 222
Auchinlek, G., 87
Audit, 63
Augmented reality (AR), 34
  glasses, 77
Autonomous robots, 46
AWAK Peritoneal Dialysis Device, 220

Barret Medical, 95
Bates, D., 181
Batteries, 83–84
  change protocol, 159–160
Berkshire Hathaway, 6
Big data management, 32
Biomed CMMS, 159
Biomedical Engineering, 31, 147, 160
Black box syndrome, 179
BLE. *See* Bluetooth Low Energy
Blood banks, 41–42
Blood glucose measurement, in wearable devices, 82–83
Blood oxygen measurement, in wearable devices, 79–80
Blood pressure measurement, in wearable devices, 81–82
Blue Ocean, 5–6
Blue Ocean Strategy (Kim and Mauborgne), 5–6
BlueBionics, 100
Bluetooth Low Energy (BLE), 84
Body temperature measurement, in wearable devices, 77–78
Brigham Health Home Hospital (MA), Huntsman Cancer Institute (UT)
  Hospital at Home program, 220–221
Buildings
  consideration with COVID-19 pandemic, 46, *47*
BURT, 95
Burton, J., 215
Butterfly IQ device, 220

Cala Health, 76
Capsix Robotics, 96
Cardiac monitoring
  patches on chest for, 74
  watches for, 74–75
Cardiac monitors, implanted, 75
Care model execution, 221–223
Caregiver, 44
CARES Act of 2020, 224
CDS. *See* Clinical decision support
Centers for Disease Control (CDC)
  on N-95 masks, 205
  on room cleaning, 206
Centers for Medicare and Medicaid Services (CMS), 13, 224
  Acute Hospital Care At Home program, 216, 221, 227
CGM. *See* Continuous Glucose Monitors
Chatbots, 175, 185
Chemical disinfection *versus* UVC, 199–200
Chiang, I., 218
Cigna, 218
Circular polarized RFID antennas, 119, *120*
Circular polarized RFID tags, 140–141
Clinical data quality, 181–183
Clinical decision support (CDS), 181, 184
Clinician assistance, of robotics
  pain relief, 96
  patient transport, 96–97
  rehabilitation robots, 95–96
Clorox, 100
CMDB. *See* Configuration management database
CMMS. *See* Computerized Maintenance Management Systems
CMS. *See* Centers for Medicare and Medicaid Services
Companions, 99
Complex medical data systems, *19*, 19–20
Computerized Maintenance Management Systems (CMMS), 55, 59
Configuration management database (CMDB), 55, 59

Connected Living, 99
Connectivity, 22–23
Consignment inventory, 135
Consolidated Appropriations Act of 2021, 225
Consumerization of health, 6
Contessa Health, 222
Continuous Glucose Monitors (CGM), 73, 74, 82
Cortana, 2
COVID-19 pandemic, 13, 101, 163, 171, 202–203, 217, 230
  building consideration with, 47–48, *47*
  contact tracing, 167–168, *168*
  hand hygiene, 202–204, 209–210
  impact on data analytics, 184–185
  impact on hospital at home, 223
  IoHT and, 44–48
  masks, 205–206
  room cleaning, 206–208
  staff, 208
Critical systems, monitoring, 134
Current Health, 219
CyberKnife system, 92
Cybersecurity, 32–33, 52
  IoHT and, 50–51
Cystic fibrosis, 245–247, *247*
Cystic Fibrosis Foundation Guidelines, 246
Cystic Fibrosis Foundation Registry, 245

Data dumping, 226
Data integrity, 31
Data issues, 178–179
Data management, using RFID, 134
Day, B., 87
Defibrillators, 42
Deployment, 66
Device monitoring, 42
Device tracking, 26
Device visualization, 26
Dexcom, 73
Diagnosis Related Groupers (DRGs), 172–174
DIEGO, 95
Digital operating room, 230–244
  building blocks, 236
  challenges in, 233–234
  challenges in pre-pandemic period, 240, *240*
  complexity of, 238–239, *239*
  critical pain points, 235–238, *237*
  future trends of, 241–243, *243*
  hospital challenges, *231*
  medical device business model, 232–233, *233*
  objectives of, *235*
  process, 238–239, *239*
  as profit center, 234–235
  status of, 239–240
  strategic imperatives for growth, *244*
  surgical volumes, *237*
  transformation, 231–232
Digital transformation, 4–5
Double Robot, 102
DRGs. *See* Diagnosis Related Groupers (DRGs)

ECG measurement, in wearable devices, 80, 83
802.11 Standard, 146
Ekso Bionics, 98
Electroencephalogram (EEG)
  measurement, in wearable devices, 80
  monitoring, 77
Electromyogram (EMG)
  measurement, in wearable devices, 80–81
Electronic Medical Record (EMR), 20, 143, 176, 181, 182
Electronic Product Code (EPC)
  Gen-II tags, 113
EMG. *See* Electromyogram (EMG)
EMR. *See* Electronic Medical Record
Energy harvesting, 84
Environmental Services (EVS), 206, 207
EPA, 198, 199
EPC. *See* Electronic Product Code
EVS. *See* Environmental Services
Exoskeletons, 97–98

FaceTime, 2
Family of robots, 92
Fanos, J., 245
FDA. *See* Food and Drug Administration

Fee for service, 171–172
Find Care Now, 5
5G
IoHT and, 49–50, *50*
Fixed RFID readers, 116, *116*
Food and Drug Administration (FDA), 73, 74, 181, 198, 218–219
on hand hygiene, 203–205
on hospital at home, 218
510K process, 53

Gaumard Advanced CPR, 103
Genome-Wide Association Studies (GWAS), 184
Genomics, 184
Germicidal ultraviolet "C" (UVC) light, 187–200
absorption, 197
angle of incidence, *194*, 194–195
canyon wall effect, 195–197, *196*
versus chemical disinfection, 199–200
current technologies, 189
definition of, 187–188, *188*
distance to target, 193, *193*
effects and mitigation on hospital surfaces, 192–193
effects and mitigation on humans, 192
emitter configurations, 197–198
lack of standards and regulations, consequences of, 198–199
limitations of, 198
line of Sight, 195
measurement devices, 189–190, *190*
poor reflection, 195
safety, 192–193
shadows, 195
susceptibilities of HAI pathogens, 190–192, *191*
transmission, 197
units of measurement, 189–190
Glucose monitoring, patches for, 73
Google, 176
Governance, 67–68
Grant audit device verification, 150–151
GSR sensor, 81
GWAS. *See* Genome-Wide Association Studies

HAIs. *See* Healthcare-associated infections; Hospital-acquired infections
Hand hygiene, 203–205, 209–210
HandX robotic instrument, 93
Hazardous contact, of robotics, 102
HDO. *See* Healthcare delivery organizations
Health Care Innovation Award, 222
Health Recovery Solutions, 223–224
Healthcare and Public Health Sector Coordinating Council (HPHSCC), 68
Healthcare-associated infections (HAIs), 199
pathogens, UVC susceptibilities of, 190–191, *191*
Healthcare delivery organizations (HDO)
medical device security program for, 52–72
HealthTech, 74
Hearables, 76
Heart monitors, 42
Heart rate measurement, in wearable devices, 79
Heinlein, R. A., 92
High-frequency (HF) RFID tags, *122*, 127
Hodivala, C., 104
Holter monitors, 74
Hospital-acquired infections (HAIs), 100, 209
Hospital at home, 213–227
care model execution, 221–223
COVID-19 pandemic, impact of, 223–225
digital solutions, 219–220
drivers for change, 214–215
future challenges to, 225–226
systemic challenges, overcoming, 220–223
technology, 217–220
traction in academia, finding, 215–217
Veterans Administration, role of, 214–215
Hospital intelligence, drivers for, 13–14
Hospital workflow, improving, 138
HPHSCC. *See* Healthcare and Public Health Sector Coordinating Council
Human Xtensions Ltd., 93
Hyland, L. A., 113

IBM Watson, 174–175
IHA. *See* Intelligent Health Association
Imaging, 176

In hospital monitoring, 75
Infant safety, enhancement of, 137–138
Information-sharing analysis centers (ISAC), 65
Infrastructure optimization, 10
Infusion systems, 153–154, *154*
  utilization, 153–154, *155*
Institutional location system, design architecture of, *25*
Insulin pumps, 74
Integration, 22–23
  of technologies and patient care environment, 16–17
Intelligent Health Association (IHA), 9
Intelligent health consumer, rise of, 1–10
Intelligent Health System. *See also individual entries*
  characteristics of, 3–4
  components of, 17–19
  objectives of, 14–16
  rise of, 3–8
Intermountain Healthcare, Utah, 221
Internet of Health Things (IoHT), 9–10
  and AI, 48–50
  blood banks, 41–42
  caregiver, 44
  and COVID-19 pandemic, 44–48
  and cybersecurity, 50–51
  device monitoring, 42
  devices, 40
  and 5G, 49–50, *50*
  in healthcare, evolution of, 40
  Medical adherence, 40–41
  remote care, 43
  remote patient monitoring, 42–43
  telemedicine, 43, *44*
  use cases, 48–49
  visibility in healthcare with, 38–51
  and wellness, 48
Internet of Things (IoT), 14, 18, 20
  and healthcare, 39–40
  mobile IoT devices, 21
  platform, 38–39
Interoperability, 22–23
InTouch, 102
Intuitive da Vinci, 92
Inventory
  consignment, 135
  control, 26
  management, 25–26, 133
IoHT. *See* Internet of Health Things
IoT. *See* Internet of Things
ISAC. *See* Information-sharing analysis centers
IUVA, 198
IV formulary, validation of, 155–156
IV pump, 66
iYU robot, 96

Jackson, S., 102
John A. Hartford Foundation, 215–216
Joint Commission, 143, 151
JP Morgan Chase, 6

Karjian, R., 222
Kim, W. C.
  Blue Ocean Strategy, 5–6
Kuka, 96

Laboratories, robotics in, 105
Laerdal, 107
LBR Med, 96
Leff, B., 215–216
Legacy analytics, in healthcare, 171–173
Lifeline Medical Alert system, 216
Linear polarized RFID antennas, 119, *119*
Linear polarized RFID tags, 140–141
Liverpool Women's Hospital, 101
Logistics, 100–101
Low-frequency (LF) RFID tags, *114*, *118*

Magnetic field
  on head to stop migraines, 77
Malins, J., 245
Marasco, P., 99
Mark, N., 221
Masks, 205–206
Massage Robotics Inc., 96
Mauborgne, R.
  Blue Ocean Strategy, 5–6
McEwen, J., 87
MDRO. *See* Multidrug-resistant organisms

Medical adherence, 40–41
Medical delivery drone, 45, *46*
Medical device business model, 232–233, *233*
Medical device security program, for healthcare delivery organization, 52–72, *56*
  asset inventory and management, 55–59
  audit, 63
  controls, *61*
  deployment, 66
  governance, 67–68
  incident detection, 64–66
  measures for success, 68–69
  mitigation, 64–66
  mobile medical devices, 69, 71–72
  network architecture, 63–64, *64*
  patch management, 62–63
  procurement, 59–60
  response, 64–66
  risk management, 60–61
  ROI indicators, 69, **70–71**, *71*
  secure configuration, 66
  segmentation, 63–64
  training, 68–69
  vendor management, 68
  vulnerability, 62–63
Medical devices
  lifecyle, *54*
  real-time tracking of, 137
Medicare, 172
Medication errors, reduction of, 138–139
MEDIport robot, 96
MedStar Washington, 100–101
Medtronic Mazor, 92
Memorial Sloan Kettering Cancer Center (MSK), 147, 148, 156, 158
  RFID clinical application pilot studies, 148–149
Metro Health, 223–224
Microneedles, 82
Microrobots, 93–94
Microsoft, 176
Mitigation, 64–66
Mobile IoT devices, 21
Mobile medical devices, 69, 71–72
Mobile or handheld RFID readers, 114–115, *115*
Motion measurement, in wearable devices, 78–79
Motorized laparoscopy, 93
Mount Sinai Health System (NY), Massachusetts General Hospital (MA), 221, 225
  Visiting Doctor's Program, 222
MSK. *See* Memorial Sloan Kettering Cancer Center
Multidrug-resistant organisms (MDRO), 188

Narrow Band, 84
National Institutes of Health (NIH), 180
Natural Language Processing (NLP), 178
Navigation assistance, of robotics, 101
Near Field Communication (NFC), 84–85
Netflix, 6
Network, 20
  architecture, 63–64, *64*
  impact on RFID/RTLS enabled applications, 147–148
Networked RFID readers, 115, *116*
NFC. *See* Near Field Communication (NFC)
NIH. *See* National Institutes of Health
NLP. *See* Natural Language Processing
N-95 masks, 205

Obamacare. *See* Affordable Care Act (ACA)
Ochsner Health System, 74
Omni-directional RFID antennas, 120, *121*
Omnicell, 104
Orbel, 209–210
  compliance at hospitals, increasing, 210–211
  distribution, 211–212
  hospitality, 211
  restaurants, 211
  retail, 211

Pacemakers, 42
Pain relief, 96
Pappas, H. P., 9–10
PARO Therapeutic Robot, 99
Passive RFID readers, 117–118, *118*

Passive RFID tags, 123, *124*
Patch management, 62–63
Patch RFID antennas, 120, *121*
Patient assistance, of robotics
  companions, 99
  exoskeletons, 97–98
  prostheses, 98–99
Patient-centric identification and association, 21–22, *22*
Patient interactive data and experience, *30*
Patient satisfaction, 156–158, *157*
Patient transport, 96–97
People issues, 179–180
Personal protective equipment (PPE), 163, 223
Personnel management, RFID in, 135
Pharmacy, robotics in, 104–105
Phased array antennas, 129–130
Phased array RFID antennas, 130, *131*
  asset management, 131
  in healthcare, 130
  patient tracking and identification, 131
  staff tracking, 131–132
Phenome Wide Association Studies (PheWAS), 184
PheWAS. *See* Phenome Wide Association Studies
Philips, 100
Philips Electronics, 113
Physical waiting room, elimination of, 165
Physiological monitoring, 19–20
Physiological sensors, in wearable sensors
  blood glucose measurement, 82–83
  blood oxygen measurement, 79–80
  blood pressure measurement, 81–82
  body temperature measurement, 77–78
  ECG measurement, 80–81
  EEG measurement, 80–81
  EMG measurement, 80–81
  heart rate measurement, 79
  motion measurement, 78–79
  respiration rate measurement, 81
PM. *See* Preventative maintenance
PPE. *See* Personal protective equipment
PPG. *See* Pulse plethysmograph sensor
Predictive analytics, 176–177
  barriers to implementing, 177, **178**
Premature Anne, 103
Presbyterian Healthcare Services (NM), 221
Preventative maintenance (PM), 136
  compliance, 151–152
Process issues, 181
Process validation, 28
Process verification, 28
Procurement, 59–60
Programmable automata, 92–93
prostheses, 98–99
Pulse plethysmograph (PPG) sensor
  for blood oxygen measurement, 79–80
  for blood pressure measurement, 81–82
  placement, *83*
Pulse Transit Time, 82
PUMA, 560, 87
PyrAmes Inc., 81–82

Quality control, RFID in, 134

Radio-frequency identification (RFID), 112–142
  analytics, use of, 137
  antennas. *See* RFID antennas
  asset management, 136
  asset tagging, 136
  asset tracing, 133–134
  asset tracking, 133–134, 136
  asset visualization, 26
  budget, 140
  clinical application pilot studies, 148–148
  components of, 114
  consignment inventory, 135
  critical systems, monitoring, 134
  data management, 134
  functionality of, 116–117
  gain, 141
  history of, 112–113
  hospital workflow, improving, 137
  infant safety, enhancement of, 137–138
  institutional implementation, *150*
  inventory management, 133
  level of security, 140
  loss prevention, 136
  management and maintenance methodology, 158, **157**

medical devices, real-time tracking of, 137
medication errors, reduction of, 138–139
passive identification tags, *27*
patient identification, 131
patient safety, 132–133, 137
personnel management, 135
polarization, 140–141
preventative maintenance, 136
quality control, 134
rapid device recalls, 135
readers. *See* RFID readers
right technology, selection of, 139
space constraints, 140
sunsetting standards, 135
tags. *See* RFID tags
temperature for supply storage, monitoring, 139
tracing, 139–140
tracking, 139–140
Read-only RFID tags, 128
Read/write RFID tags, 129
Real-time location systems (RTLS), 24
ambulatory care, 164
asset visualization, 26
device tracking and visualization, 26
workflow metrics and optimization, 29–30
Regulatory inventory validation, 151–152
Rehabilitation robots, 95–96
Remote care, 43
Remote patient monitoring, 42–43
Respiration rate measurement, in wearable devices, 81
Return on investment (ROI), 144, 146
indicators, 69–70, **70–71**, *71*
ReWalk exoskeleton, 98
RFID. *See* Radio-frequency identification
RFID antennas
circular polarized, 119, *120*
linear polarized, 119, *119*
omni-directional, 120, *121*
patch, 119, *119*
phased array. *See* Phased array RFID antennas
RFID readers
active, 116–117, *117*
fixed, 115, *116*
mobile or handheld, 114–115, *115*
networked, 115, *116*
passive, 117–118, *118*
semi-active, 118–119
RFID/RTLS enabled applications, 24–25, 143–160
alternative technologies, 146–147
application development, 160
application integrity, 159
battery change protocol, 159–160
critical device tracking and availability, 152–153
design, 147
evaluation, 147
existing infrastructure, leveraging of, 146–147
grant audit device verification, 150–151
infusion system utilization, 154–155, *154*
IV formulary, validation of, 155–156
IV infusion systems, 153–154, *154*
network impact, 147–148
patient satisfaction, 156–158, *157*
pilot studies, 147
PM compliance, 151–152
pre-deployment analysis, 147
regulatory inventory validation, 151–152
reporting, 160
safety workflow optimization, 156–158, *157*
strategic planning and design, 145–146
tag management, 159
use cases, 150–158
RFID tags, *122*
active, 123–125
circular polarized, 140–141
frequency bands, 126, **126**
high-frequency, *122*, 127
linear polarized, 140–141
low-frequency, *122*, 122
management, 159
memory capacity, 127
passive, 123–125, *124*
read-only, 128
read/write, 129
semi-active, 125
ultra-high-frequency, *122*, 127
write-once/read-many (WORM), 128–129

Risk management, 60–61
Robotic wheelchairs, 97
Robotics, 45, 87–109
  adoption factors of, 104, *105*, *106*
  advantages of, *106*, 106–107
  autonomous, 46
  clinician assistance, 95–97
  disadvantages of, 106–107
  family of, 92
  future direction of, *108*, 108, *108*
  healthcare applications of, 87, *87*
  industrial automation, 104–105
  mobile and remote services, 99–102
  navigation assistance of, 101
  patient assistance, 97–99
  surgeon assistance, 80–95, *90*
  technologies, 88, *89*
  training, 102–103
Rock Health consumer survey, 224
ROI. *See* Return on investment
Room cleaning, 206–208
Royal Philips, 219
RTLS. *See* Real-time location systems

Safety workflow optimization, 156–158, *157*
Sam Club, 6
Samsung Electronics Co. Ltd
  Galaxy Watch3, 75
SARS-CoV-2, 187–189
SBRT. *See* Stereotactic body radiation therapy
SCFCC. *See* Southern California Cystic Fibrosis Collaborative
SDoH. *See* Social determinants of health (SDoH)
Secure configuration, 66
Segmentation, 63–64
SeizeIT, 43
Semi-active RFID readers, 118–119
Semmelweis, I., 202
Sensors, 17–19
SimJunior, 103
SimMan, 103
SimMom, 103
SimNewB, 103
Siri, 2, 176
Sleep monitors, 76
SMAC (social, mobile, analytics, and cloud) technologies, 4, 5
Smart manikins, 103
Social determinants of health (SDoH), 174, 176
Southern California Cystic Fibrosis Collaborative (SCFCC), 246
SRS. *See* Stereotactic radiosurgery
Stereotactic body radiation therapy (SBRT), 93
Stereotactic radiosurgery (SRS), 93
Sterilization, 100
Stimulation, 76
Stockman, H., 113
Strategic planning and design, 145–146
Streamlined door-to-exam patient flow, 165, *166*
Supply chain, 25–26
Surge pricing, 1
Surgeon assistance, of robotics, 90–95, *90*
  AI guided, 94–95
  assistive guides, 91–92
  microrobots, 93–94
  motorized laparoscopy, 93
  programmable automata, 92–93
  types of, *91*
  Waldos, 92
Surgical 3D printing applications, 34–36, *35*, *36*

Target, 113
Taylor, A. H., 113
TCO. *See* Total cost of operation
Technologies and patient care environment, integration of, 16–17
Technology asset, 30–32
Technology management, 30–32
Techquity, 245–248, *247*
Teladoc, 102
Telehealth, *44*, *45*
Telemedicine, 43, *44*
  robots, 101–102
Telerobots, 46
Temi robot, 99
TENS. *See* Transcutaneous Electrical Nerve Stimulation

Texas Medical Center (TMC) Innovation Institute, Houston, 105
The Joint Commission (TJC), 13, 210
Thera-grippers, 94
Theremin, L., 112
Thoracic impedance, 81
3D printing, 34–35, *35*, *36*
3D technologies
  patient-specific clinical treatment solutions, 33–34
  surgical 3D printing applications, 34–36, *35*, *36*
Threshold Limit Value (TLV), 192
Time synchronization, 20–21
TJC. *See* The Joint Commission
TLV. *See* Threshold Limit Value
Total cost of operation (TCO), 40
Transcutaneous Electrical Nerve Stimulation (TENS), 76
Transducers, 17–19
TyroMotion, 95

Uber
  AI-enabled app, 1
UI. *See* User Interface
Ultra-high-frequency (UHF) RFID tags, *122*, 127
Ultraviolet germicidal irradiation (UVGI), 187–188
UMass Memorial Hospital
  Hospital at Home program, 220
UnityPoint Health (IA), 221
Use cases
  AR and VR glasses, 77
  chest patches for cardiac monitoring, 74–75
  EEG monitoring, 77
  glucose monitoring, patches for, 73
  hearables, 76
  in hospital monitoring, 75
  implanted cardiac monitors, 75
  insulin pumps, 74
  IoHT, 48–49
  magnetic field on head to stop migraines, 77
  RFID/RTLS enabled applications, 149–151
  sleep monitors, 76
  stimulation, 76
  watches for cardiac monitoring, 74–75
User Interface (UI), 7
Utilization statistics, 28
UVC. *See* Germicidal ultraviolet "C" light
UVGI. *See* Ultraviolet germicidal irradiation

Veterans Administration (VA), 220
  Hospital-Based Home Care (HBHC) program, 214
Vgo, 102
Virtual Local Area Networks (VLANs), 63
Virtual reality (VR), 35
  glasses, 77
Virtual waiting rooms, creation of, 165
VitalPatch, 219
VLANs. *See* Virtual Local Area Networks
Voice recognition, 175–176
VR. *See* Virtual reality
Vulnerability, 62–63

Walgreens, 5
Walmart, 5–6, 113
Walton, C. A., 113
Watson-Watt, Sir R., 113
Wearable devices
  physiological sensors in. *See* Physiological sensors, in wearable sensors
  wireless communication for, 84–85, *85*
Weaver, S., 98
Wellness, IoHT and, 48
WHO. *See* World Health Organization
WiFi, 84
Wikipedia, 6
Windows XP, 62
Wireless communication, for wearable devices, 84–85, *85*
Wireless infrastructure, 20
Wireless standards, 84, *85*
Wisepill, 41
Wolfberg, A., 225
Workflow metrics, 29–30
Workflow optimization, 29–30
Workflow^{RT}, 165, 166, *166*
  contract tracing report, 168, *168*
  staying connected to families and visitors, 168–169

World Health Organization (WHO), 40
on hand hygiene, 203, 209
WORM. *See* Write-once/read-many (WORM) RFID tags
Write-once/read-many (WORM) RFID tags, 128–129

Xenex LightStrike robot, 100
XR. *See* Augmented reality (AR); Virtual reality (VR)

Yelp, 5
Young, L. C., 113

Printed in the United States
by Baker & Taylor Publisher Services